MEMBRANES AND MOLECULAR ASSEMBLIES:
THE SYNKINETIC APPROACH

Monographs in Supramolecular Chemistry
Series Editor: J. Fraser Stoddart, FRS
University of Birmingham, UK

This series has been designed to reveal the challenges, rewards, fascination, and excitement in this new branch of molecular science to a wide audience and to popularize it among the scientific community at large.

No. 1 Calixarenes
By C. David Gutsche, Washington University, St. Louis, USA

No. 2 Cyclophanes
By François Diederich, University of California at Los Angeles, USA

No. 3 Crown Ethers and Cryptands
By George W. Gokel, University of Miami, USA

No. 4 Container Molecules and Their Guests
By Donald J. Cram and Jane M. Cram, University of California at Los Angeles, USA

No. 5 Membranes and Molecular Assemblies: The Synkinetic Approach
By Jürgen-Hinrich Fuhrhop and Jürgen Köning, Freie Universität Berlin, Germany

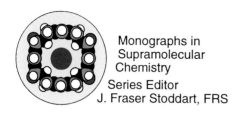 Monographs in
Supramolecular
Chemistry
Series Editor
J. Fraser Stoddart, FRS

Membranes and Molecular Assemblies:
The Synkinetic Approach

Jürgen-Hinrich Fuhrhop and Jürgen Köning
Freie Universität Berlin, Germany

THE ROYAL
SOCIETY OF
CHEMISTRY

The cover diagram shows a pore in a vesicle membrane which allows the flow of lithium ions. It is sealed here with a long-chain "stopper". See Figure 4.24 on page 82.

ISBN 0–85186–732–4 (Hardback)
0–85186–714–6 (Softback)

A catalogue record of this book is available from the British Library

© The Royal Society of Chemistry 1994

All Rights Reserved
No part of this book may be reproduced or transmitted in any form or by any means – graphic, electronic, including photocopying, recording, taping, or information storage and retrieval systems – without written permission from The Royal Society of Chemistry

Published by The Royal Society of Chemistry,
Thomas Graham House, Science Park, Milton Road, Cambridge CB4 4WF, UK

Colour artwork on disk processed by Goodfellow and Egan Ltd, Cambridge, UK
Typeset by Computape (Pickering) Ltd, Pickering, North Yorkshire, UK
Printed and bound by Cambridge University Press, Cambridge, UK

Foreword

Recent developments in supramolecular chemistry are explosive. This is attested by flourishing new journals and monographs with this name and, above all, by frequent use of this word in professional journals. Supramolecular chemistry attempts to bridge wide-ranging areas of science and technology. This aspect effectively promotes creation of varied interdisciplinary fields.

The major focus of this book is membranes and molecular assemblies which occupy a central importance in supramolecular chemistry. These areas have largely been studied by physical chemists and biophysicists, and organic chemical aspects have often been ignored. This situation has changed rapidly in recent years. Organic chemists started to synthesize new types of molecules that would spontaneously assemble in specific ways. They soon realized that a variety of interesting molecular assemblies were within their reach and were closely related from the synthetic point of view.

The book by Fuhrhop and Köning provides a new vista in this respect. It deals with micelle, microemulsion, monolayer and bilayer vesicle, micellar fibre, nanopore, amphiphilic crystal, and oligomolecular assembly. In spite of the much varied physicochemical characteristics of these molecular systems, common features which justify the synkinetic approach, as the authors advocate, emerge immediately. This unified understanding should enrich supramolecular chemistry. The skillful narrative of Fuhrhop, an accomplished chemist with the flair of a novelist, is apparent, as he has demonstrated in his previous book[1]. The extensive use of computer graphics is an invaluable service to readers. I hope this book helps foster a new breed of chemists.

Toyoki Kunitake

Kyushu University, Japan
August 1994

---

[1] J. Fuhrhop, G. Penzlin, Organic Synthesis: Concepts, Methods, Starting Materials, 2nd edn, VCH, New York, **1994**.

Preface

This book has been written and computer-drawn to present the wealth of membraneous structures that have been realized by chemists mainly within the last ten years. The models for these artificial molecular assemblies are the biological lipid membranes; their ultimate purpose will presumably be the verification of vectorial reaction chains similar to biological processes. Nevertheless, chemical modelling of the non-covalent, ultrathin molecular assemblies developed quite independently of membrane biochemistry. From the very beginning of artifical membrane and domain constructions, it was tried to keep the preparative and analytical procedures as simple and straightforward as possible. This is comparable to the early development of synthetic polymers, where the knowledge about protein structures quickly gave birth to simple and more practical polyamides.

In the first chapter, the words "synkinesis" for the synthesis of non-covalent molecular assemblies and "synkinons" for the corresponding monomers are introduced for the first time. These target-structure oriented words, analogous to "synthesis" and "synthon", are felt to be more suitable than the physical terms, such as aggregation, self-organization, building blocks, surfactants, detergents etc. One should realize that "synkinetic" structures have become quite complex and that the planning of "self" organization makes little sense.

The other chapters then lead from the simple to the more complex molecular assemblies. Syntheses of simple synkinons are described at first. Micelles made of 10–100 molecules follow in chapter three. It is attempted to show how structurally ill-defined assemblies can be most useful to isolate single and pairs of molecules and that micelles may produce very dynamic reaction systems. A short introduction to covalent micelles, which actually are out of the scope of this book, as well as the discussion of rigid amphiphiles indicate where molecular assembly chemistry should aim at, namely the synkinesis of solid spherical assemblies. Chapter four dealing with vesicles concentrates on asymmetric monolayer membranes and the perforation of membranes with pores and transport systems. The regioselective dissolution of porphyrins and steroids, and some polymerization and photo reactions within vesicle membranes are also described in order to characterize dynamic assemblies.

Chapter five introduces highly organized, quasi one-dimensional crystals, namely micellar rod and vesicular tubular fibres. They are compared to equally fascinating liquid threads found in viscoelastic gels and to phospholipid tubules. These membraneous assemblies build a bridge to the secondary structures of covalent polymers and biopolymers. Chapter six is about receptor properties of monolayers on water, which can be easily detected on the Langmuir trough and on nanopores on metal surfaces. Chapter seven finally introduces 2D and 3D crystals with receptor properties and/or nanometre cavities.

The terms "molecular machines" or "molecular architecture" are avoided. They imply a complexity and efficiency which none of today's synkinetic systems can attain. It is, however, hoped that the pictures and discussions of the systems as they are known today convey the impressions of easy accessibility in excellent yields, well-defined stereochemistry of membranes and domains, as well as interesting supramolecular reactivities.

To conclude this preface, I wish to thank enthusiastic, courageous and skilled co-workers for their contributions to the development of asymmetric monolayer membranes, pores and fibrous assemblies; their names are found in references throughout this book. The work was generously supported by the Deutsche Forschungsgemeinschaft within a "Sonderforschungsbereich" called Vectorial Membrane Processes, as well as by the Freie Universität Berlin (FNK) and the Fonds der Chemischen Industrie. Mrs Regina Stück has typed the manuscript perfectly starting from illegible manuscripts, and Mrs Caroline Friedrich improved the English.

Dipl. Ing. Alexander Maack helped when the manuscript's deadline approached. He produced several of the computer illustrations. Many thanks go to my colleagues who provided an extremely friendly community in Gordon conferences and several meetings in Japan, the United States and Europe. Several of them also visited Berlin. All I learnt about membranes and molecular assemblies comes either from them or from my students. And then we have to thank Dr Anthony Breen, who took care of rapid publication and gave the book a final friendly touch.

Jürgen-Hinrich Fuhrhop

Berlin
May 1994

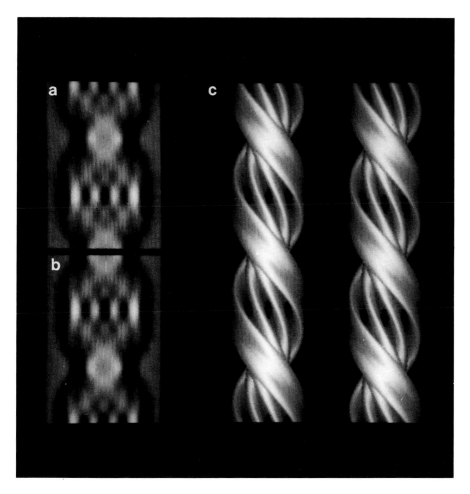

N-*Octyl*-D-*gluconamide self-assembles into long fibres in water. Analysis of electron micrographs showed four fibres entwined to form a quadruple helix (Figure 5.13 on pages 116–117). Originally, each thread was thought to have a circular cross-section of bimolecular thinness.*

However, a few weeks before this book was published, a new chapter was added to the endless story of the N-*octyl*-D-*gluconamide helix. Phosphotungstate staining combined with rapid freeze techniques and cryo transmission electron microscopy gave an extremely well resolved, non-distorted projection image (a) (shown above). This was used to produce a 3D image directly, without any assumptions. The new technique suggests the quadruple helix (b) with bilayer ribbons instead of bilayer threads (see page 116–123) as units. Reprojection (c) of the 3D model (b) shows excellent agreement with the input image (a). The bulges of the original model, which were difficult to understand, have disappeared. But now, one has replaced the simple threads of circular cross-section, which were so obvious in visual inspections of electron micrographs, by ribbons of a uniform width of 97 Å. The new structure thus presents a new problem: why does 2D growth of the crystalline bilayer stop just at 97 Å and why does visual inspection of electron micrographs not reveal ribbons? (By courtesy of Dr Chrisoph Böttcher)*

Contents

Chapter 1	Synkinesis and Synkinons of Supramolecular Assemblies	1
	1.1 Introduction	1
	1.2 Target Assemblies	2
	1.3 Synkinons	4
Chapter 2	Syntheses of Synkinons for Supramolecular Assemblies	7
Chapter 3	Micelles and Microemulsions	20
	3.1 Introduction	20
	3.2 Why Micelles are Formed and Remain in Solution	21
	3.3 Structural Models of Micelles and the Wetness Problem	24
	3.4 No Known Surfactant Has a Conical Shape	28
	3.5 The Millisecond Life of Individual Micelles and their Featureless Appearance	29
	3.6 The Dissolving of Large Molecules via Micelles – One or Two per Micelle	30
	3.7 Light-Initiated Electron Transfer in Micelles	33
	3.8 Micelles Made of Rigid Amphiphiles	34
	3.9 Conversion of Micelles to Vesicles and Thin Films	36
	3.10 The Covalent Micelle-like Behaviour of Dendritic Macromolecules	39
	3.11 The Solubilization of Enzymes, Carbohydrates and Inorganic Colloids in Organic Solvents	43
	3.12 Fat Droplets and Microemulsions and their Application in Surface Cleaning and as Catalytic Systems	44
Chapter 4	Molecular Monolayer and Bilayer Vesicle Membranes	49
	4.1 Introduction	49
	4.2 The Self-Assembly of Vesicles and why they Remain in Solution	50

4.3	Synkinesis of Asymmetric Vesicle Membranes (In–Out)	55
4.4	Membrane Dissolved Dyes and Steroids	66
4.5	Ion Transport, Domain Formation and Pores in Vesicle Membranes	75
4.6	The Entrapped Water Volume	85
4.7	Polymerization in Vesicle Membranes and Interactions with Polymers	85
4.8	Photoreactions in Vesicle Membranes	92
4.9	Rearrangements of Vesicles	93

Chapter 5 Binding Interactions in Micellar and Vesicular Fibres   98

5.1	Introduction	98
5.2	Fluid Molecular Bilayer Scrolls	99
5.3	Fluid Micellar Threads	101
5.4	Amide Hydrogen Bond Chains	104
5.5	Solid Molecular Mono- and Bilayer Twisted Ribbons and Tubules	106
5.6	Molecular Mono- and Bilayer Micellar Rods	114
5.7	Co-Crystallized Micellar Fibres	123
5.8	Porphyrin and Metalloporphyrin Fibres	126
5.9	Biopolymer Fibres	139
5.10	Helical Metal Complexes	144
5.11	A Stereoselective Pitch Depression in a Cholesteric Phase	147

Chapter 6 Molecular Recognition and Nanopores in Surface Monolayers   149

6.1	Introduction	149
6.2	Molecular Recognition at Monolayers on the Water Surface	150
6.3	Molecular Recognition at LB Films and Self-Assembled Monolayers	159
6.4	Nanopores in Self-Assembled Monolayers	165
6.5	Polymers, Cast Films and Multilayers	170
6.6	Bolaamphiphiles and Dyes on Solid Surfaces	175

Chapter 7 Amphiphilic Crystals and Hydrogen Bonded Co-Crystals   182

7.1	Introduction	182
7.2	Chiral and Crystalline Langmuir Monolayers on Water	182
7.3	Solubilities and Single Crystals of Achiral Amphiphiles	185
7.4	Single Crystals of Chiral Amphiphiles	193

	7.5	Single Crystals of Carotenoids and Porphyrins	200
	7.6	Single Crystals and Fibres of Amphiphilic Steroids	203
	7.7	Molecular Recognition at Crystal Interfaces	204
	7.8	Crystal and Co-Crystal Engineering by Hydrogen Bonding and by the Filling of Voids	207
	7.9	The Future of Synkinetic Membrane and Molecular Assembly Chemistry	213

Subject Index   214

CHAPTER 1
Synkinesis and Synkinons of Supramolecular Assemblies

1.1 Introduction

Biological cell membranes are multi-component systems consisting of a fluid bilayer lipid membrane (BLM) and integrated membrane proteins. The main structural features of the BLMs are determined by a wide variety of amphiphilic lipids whose polar head groups are exposed to water while hydrocarbon tails form the nonpolar interior. The BLMs act as the medium for biochemical vectorial membrane processes such as photosynthesis, respiration and active ion transport. However, they do not participate in the corresponding chemical reactions which occur in membrane-dissolved proteins and often need redox-active cofactors. BLMs were therefore mostly investigated by physical chemists who studied their thermodynamics and kinetic behaviour[1-6].

Up until 1977, the non-covalent polymeric assemblies found in biological membranes rarely attracted any interest in supramolecular organic chemistry. Pure phospholipids and glycolipids were only synthesized for biophysical chemists who required pure preparations of uniform vesicles, in order to investigate phase transitions, membrane stability and leakiness, and some other physical properties. Only very few attempts were made to deviate from natural membrane lipids and to develop defined artificial membrane systems. In 1977, T. Kunitake published a paper on "A Totally Synthetic Bilayer Membrane" in which didodecyl dimethylammonium bromide was shown to form stable vesicles[7]. This opened the way to simple and modifiable membrane structures. Since then, organic chemists have prepared numerous monolayer and bilayer membrane structures with hitherto unknown properties and coupled them with redox-active dyes, porous domains and chiral surfaces. Recently, fluid bilayers found in spherical vesicles have also been complemented by crystalline mono-

---

[1] J.H. Fendler, Membrane Mimetic Chemistry, Wiley, New York, **1982**
[2] J.N. Israelachvili, Intermolecular and Surface Forces, Academic Press, London, **1985**
[3] G. Cevc, D. Marsh, Phospholipid Bilayers, Wiley, New York, **1987**
[4] A. Ulman, Ultrathin Organic Films, Academic Press, Boston, **1991**
[5] J.H. Clint, Surfactant Aggregation, Blackie, Glasgow, **1992**
[6] M. Shinitzky (ed.), Biomembranes, Physical Aspects, VCH, Weinheim, **1993**
[7] T. Kunitake, Y. Okahata, *J. Am. Chem. Soc.*, **1977**, *99*, 3860

and bilayers in fibrous molecular assemblies. The structural chemistry of non-covalent polymeric assemblies has slowly become as colourful as the structural chemistry of covalent amphiphilic polymers, such as proteins and nucleic acids, which dominate reaction patterns in biological chemistry.

1.2 Target Assemblies

This book deals with the organic chemistry of micelles, vesicles, micellar fibres, surface monolayers and a few 3D crystals formed by the assembly of synthetic surfactants in water or, less common, in organic solvents. Common features of these assemblies are molecular thinness and direct interaction of all their molecules with the environment, usually aqueous media.

Physical chemists established a process called "self-organization" in which water-insoluble amphiphiles firstly form a molecular brush on the water surface and then assemble to spherical droplets or "bladders" in bulk water if a threshold concentration (cmc, critical micellar concentration) is surpassed. It was also shown that the self-organization of molecular mono- and bilayers is commonly not followed by crystal growth which would normally be favoured as it diminishes surface energies. Repulsive hydration and undulation effects were held responsible for preventing the growth of the delicate bilayer structures to 3D crystals.

Self-organization, however, is unsatisfactory for producing assemblies with a defined stereochemistry and functionality or with nanopores and receptors which are characteristic of biological membranes. These features have to be planned by organic chemists who also define single or multi-component molecular assemblies. We therefore propose here the word **"synkinesis" for the synthesis of a non-covalent molecular assembly** with a defined structure and/or function. **"Synkinons"** is then the name of the **building blocks** of such target assemblies. Membranes and molecular assemblies do not "self-organize", but follow the "synkinetic" plans of a chemist. The new term should also express speed. Ultrathin assemblies are often formed within seconds, whereas synthetic reactions and crystallization are typically slow. Furthermore "kinesis" or motion does usually not only occur in one direction. Individual molecules can always migrate away or "dissociate" from a molecular complex, membrane or assembly. Here lies a clear distinction to synthetic molecules, where activation energies are high in both the direction of synthesis and decomposition.

We substantiate the new notions of synkineses and synkinons for several important types of molecular assemblies as described in this book.

Micelles are short-lived, ill-defined assemblies (see chapter 3) of 50–100 amphiphilic molecules in water. They can be target assemblies of synkineses if they answer a well-defined purpose. For example, one may design a system in which only two different molecules form a heterodimer within each micelle. In this case, the micelle could be a very simple SDS-micelle, but the components of the dimer must be carefully fitted to the micelle's size and lifetime. Synkinesis would concentrate on the structure of the dimer components rather than the micelle. Micelles may also be used in an irreversible light-induced charge transfer process. Here, the educts and products of a photoreduction must be synchronized with the properties of the micelle in order to allow charge separation and to slow down charge recombination. Synkinetic planning is again a presupposition of a successful system. Or one may wish to produce a spherical micellar particle with fixed boundaries and a long lifetime. In this case, the hydrated head groups which repulse each other must be replaced by head groups which form strong hydrogen bonds and possess a concave shape. In all three cases, the "synkinons" are not only forced together by the hydrophobic effect, but are designed as building blocks of a designed molecular assembly or reaction system.

Vesicles which are long-lived and chromatographable (see chapter 4), are even more appropriate as target assemblies of synkinetic planning. The preparation of asymmetric vesicle membranes in which the outer and inner surfaces act as electron donors and acceptors, or vice versa, are of high relevance. A membrane of this type may support light-induced charge separation. The reversible perforation of a vesicle membrane with pores needs careful synkinetic planning and cannot simply rely on self-organization of a single amphiphile. This is also in keeping for photoactive membrane transport systems and vesicles which dissolve heme analogues and transport molecular oxygen. Since synkinetic procedures are generally reversible, vesicles can form buds or can fuse. Both processes can be induced by a change of environmental conditions or can be completely inhibited by covalent polymerization of appropriate synkinons.

The most complicated and structurally best-defined target assemblies of synkinesis are **micellar or vesicular fibres**. Their structure is fixated by bonding interactions along the fibre axis. The width of the fibre as well as the pitch of eventual helices can be reproduced within a range of a few Ångströms. Most of these well-defined fibres can actually be envisioned as one-dimensional crystals; some may be connected to tissue-like cloths. One may also functionalize these fibres by selecting redox- and photo-active porphyrins as hydrophobic cores. Furthermore, vesicular tubules can be combined with inorganic coatings or entrapments by a sequence of synkinetic steps to form ultrathin ionic or metallic nano-wires.

Surface monolayers on water, metals, silica or organic polymers do also not only "self-organize" or "self-assemble", but they can be selectively functionalized as well with dyes, stereoselective receptors or nanoholes. Several monolayers can also be synkinetized subsequently to form redox chains with different electron donors or acceptors in multilayers. Vectorial reaction chains can thus be materialized in synkinetic materials.

Finally, the **surface layers of 3D crystals** can also be employed as templates for the growth of organized multilayers or new crystal structures. Reactive *co-crystals*, or crystals with large *cavities* or inner surface areas have also been designed. Again simple "self-organization" of molecules in bulk solution simply leads to more or less homogeneous crystals with no chemical activity. Only synkinesis of assemblies containing intelligent combinations of synkinons gives systems with new material properties.

Comparisons between the synthesis of covalent molecules and the synkinesis of non-covalent molecular assemblies immediately reveal the more "fugitive" or reversible character of the latter. Although some molecular assemblies can be isolated in solid form and even be crystallized, they are normally only detectable, stable and active in bulk aqueous media. This is characteristic and in common with biological materials or "organs". The purpose of this book is to show how synkinetic molecular assemblies can be formed and their partial likeness to the structures and functionality observed in biological structures. Synkinesis has now been pursued for 17 years and Kunitake's DODAB vesicle, which has been taken as the starting point, has triggered the development of an enormous variety of beautiful supramolecular architecture.

1.3 Synkinons

An electron donor or acceptor site is usually needed in organic synthons for covalent synthesis. The covalent connection of both leads to a new molecule in an essentially irreversible synthetic reaction. Organic synkinons for non-covalent synkinesis usually contain a hydrophilic and a hydrophobic part and/or proton donor or acceptor sites. Non-covalent connection of such amphiphiles leads to a molecular assembly in a reversible synkinetic reaction. **Amphiphiles are not only surface active molecules ("surfactant, detergent"), but much more important, they *create* surfaces**. This becomes particularly evident in microemulsions and in suspensions of vesicles and micellar fibres, but is also true in nanoholes and pores, on monolayer surfaces and for many other supramolecular structures.

There are, for example synkinons for the synkinesis of micelles, vesicles, pores, fibres and planar mono- or multilayers. A given synkinon can also be applied for another synkinetic target if the conditions are changed or if the synkinon is chemically modified. The most simple example is stearic acid. At pH 9, it is relatively well-soluble in water and forms spherical micelles. If provided with a hydrogen bonding chiral centre in the hydrophobic chains (12-hydroxystearic acid), it does not only form spherical micelles in water but also assembles into helical fibres in toluene. At pH 4, stearic acid becomes water-insoluble but does not immediately crystallize out; spherical vesicles form. A second type of synkinon, which produces perfectly unsymmetrical vesicle membranes, consists of bolaamphiphiles with two different head groups on both ends of a hydrophobic core. Such bolaamphiphiles are also particularly suitable for the stepwise construction of planar multilayered assemblies.

Different synkinons are used to assemble domains and pores within a membrane structure. Such synkinons ideally have the length of the molecular mono- or bilayer in which they participate and they are made rigid by two parallel alkyl chains of a macrocycle or by a combination of several small rings which are connected to a chain. Such **amphiphiles** can then be forced to form a domain if they **contain a hydrophobic, membrane-soluble edge together with a hydrophilic edge** which **assemble to form hydrated pores**. Another important type of synkinon is provided by **amphiphilic or bolaamphiphilic dyes**. They may

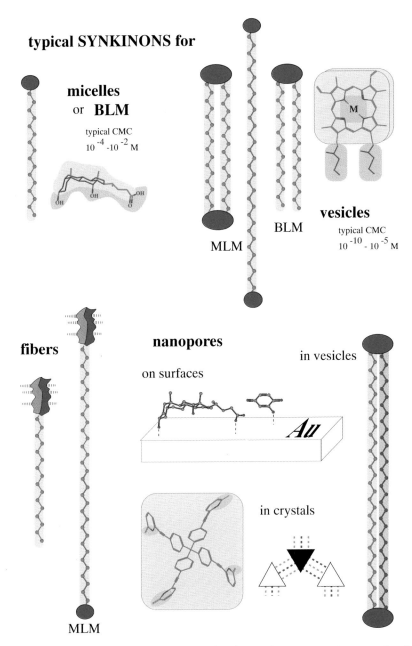

Figure 1.1 *Schematic models of a few typical synkinons for some membranes and molecular assemblies. **Green indicates hydrophobic; blue is hydrophilic**. In a few instances red and blue are applied to symbolize proton donor and acceptor sites in hydrogen bonds. Hydrophobic volumes are symbolized by green areas, hydrophilic head group regions are often depicted as blue zones. BLM means bilayer lipid membrane; an MLM is a monolayer lipid membrane.*

be linear polyenes and are already considered as electron wires in membranes or they may act as photoactive redox systems, such as quinones or porphyrins, in charge separation systems. The dissolution of any amphiphilic synkinon as monomer or domain into fluid membrane systems is normally uncomplicated. Their ordered arrangement in protein-like, crystalline fibres is, however, a largely unsolved problem.

Molecular tools such as stoppers for pores, fluorescent dyes as indicators for changes in membrane properties, water-dissolved **signal molecules or electron acceptors etc.** can be seen as final requisites for synkinetic assemblies. Synkinesis therefore does not only imply the preparation of molecular assemblies, but also the establishment of optimal environmental conditions for their eventual activities.

CHAPTER 2

Syntheses of Synkinons for Supramolecular Assemblies

Biological membranes are composed of a chemically inactive lipid bilayer and reactive proteins. Synkinetic molecular assemblies which mimic some aspects of biological structures or membrane reaction chains have no requirement for biopolymers. Instead **effective synkinons** which **possess at least one of four important structural and functional properties** of proteins are needed. These are **(i) amphiphilic character, (ii) asymmetry, (iii) secondary amide groups and (iv) redox- and photochemically active cofactors**. In a general sense, amphiphilic character **gives access to** organic media for the **dissolution of apolar molecules in bulk water**. Asymmetry allows a **molecular ordering** in which sequences of different molecular "synkinons" can identify with each other and be combined to form complex structures or reaction chains. Amide hydrogen bonds, or equivalent linear and two-sided interactions, cement these supramolecular assemblies. Dyes **interact with sunlight**, electrons and oxygen to bring power into the housings of molecular dimensions.

In this chapter we introduce compounds which have been successfully applied in the construction of supramolecular assemblies. Only the amphiphiles which have been prepared in sufficient quantities have been admitted; milligram quantities being considered unacceptable as starting materials for the preparation, analysis and application of assemblies. Experience proves that complicated dyes, pore builders, receptors etc. never reappear in the literature after their syntheses and spectroscopic properties have been reported. On the other hand, such easily attainable synkinons and surfactants around the ten gram scale need not, of course, be too simple. On the contrary, they may contain all the components of the chiral pool, i.e. amino acids, carbohydrates, steroids etc., as well as all commercial dyes of interest such as protoporphyrin, phthalocyanines, carotenes, viologen and quinones.

Furthermore, the following special requirements for an amphiphile synthesis should be taken into consideration. **Chromatography and extraction procedures should be avoided during the isolation of amphiphiles** as this type of compound is inherently extremely polar, insoluble in most solvents and produces foam. The synthesis should therefore aim for apolar precursors into which the **polar head groups can be introduced in one final high yield reaction**. Minimal amounts of impurities can then be removed by precipitation or, ideally, via crystallization procedures. Saponification of esters, Michael additions of thioglycosides to maleic esters, and alkylation of amines to form tetraalkylammonium salts are typical examples of such high-yield, one-step conversions from apolar precursors to polar amphiphiles. The reductive cleavage of disulfides to hydrosulfides, required for self-assembly on gold surfaces, also belongs to this category.

The most versatile synthetic building blocks for the syntheses of single-chain synkinons are ω-halogeno-esters, -nitriles, or -amines, which are ideal for the **introduction of deuterium**. Terminal trideuteromethyl groups were introduced with deuterated dimethyl lithium cuprate[1]. Methylene groups next to hydroxy or amino groups are routinely obtained by the action of lithium aluminium hydride on esters or nitriles[2]. Deuteration in the centre of an oligomethylene chain may involve the Wittig synthesis of an olefinic ester or nitrile and subsequent hydrogenation with homogeneous Wilkinson catalyst and deuterium[3]. Heterogeneous catalysis often leads to a scrambling of deuterium atoms (Scheme 2.1)[3].

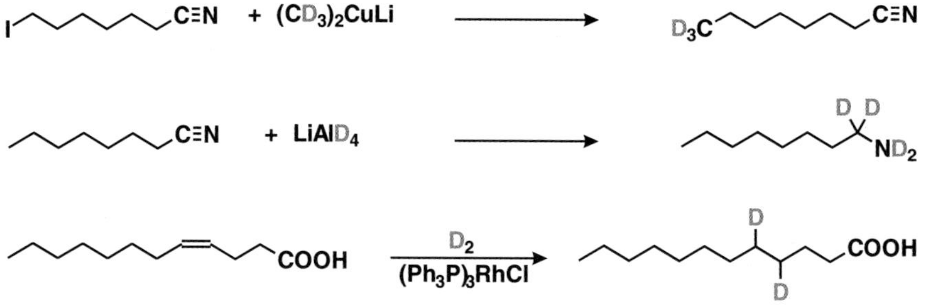

Scheme 2.1

Many glycero- and phospholipids – chemicals of major interest to physical chemists and biochemists dealing with "biomembranes" – are available commercially. A list of the most common compounds is given below (Scheme 2.2).

Total syntheses of many **glycerolipids** involving protecting groups for both the glycerol and phosphate groups have been reviewed[4]. Stereoselective syntheses of glycerolipids have also been reported[5]. Asymmetric phosphodiesters with two different alcohols were prepared from the diphosphate via the monoalkyl dihydrogenphosphate in two steps[6]. A large number of phospholipid "con-

---

[1] G.M. Whitesides, W.F. Fischer jr, J. San Filippo jr, R.W. Bashe, H.O. House, *J. Am. Chem. Soc.*, **1969**, *91*, 4871
[2] L.H. Amundsen, L.S. Nelson, *J. Am. Chem. Soc.*, **1951**, *73*, 242
[3] J.R. Morandi, H.B. Hensen, *J. Org. Chem.*, **1969**, *34*, 1889
[4] H. Eibl, *Angew. Chem.*, **1984**, *96*, 247; *Angew. Chem., Int. Ed. Engl.*, **1984**, *23*, 257
[5] G. Lin, M.-D. Tsai, *J. Am. Chem. Soc.*, **1989**, *111*, 3099
[6] a) A.L. Wagenaar, L.A.M. Rupert, J.B.F.N. Engberts, D. Hoekstra, *J. Org. Chem.*, **1989**, *54*, 2638
b) R.A. Bauman, *Synthesis*, **1974**, 870

Syntheses of Synkinons for Supramolecular Assemblies

Some Glycerophospholipids:

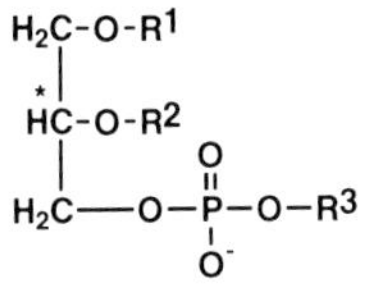

H_2C-O-R^1
$HC-O-R^2$ (*)
$H_2C-O-P(=O)(O^-)-O-R^3$

* D- or L-configuration at C-2 of glycerol

Sphingophospholipids contain a secondary amide bond:

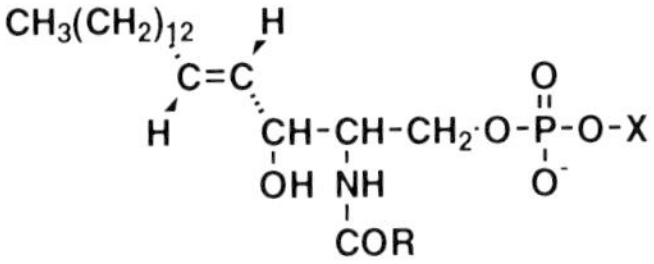

$CH_3(CH_2)_{12}-CH=CH-CH(OH)-CH(NH-COR)-CH_2-O-P(=O)(O^-)-O-X$

Name	Abbr.	R^1	R^2	R^3
Oleoyl-lysophosphatidic acid		$-OC(CH_2)_7CH=CH-C_8H_{17}$ (cis)	H	H
L-d-Dimyristoyl-phosphatidic acid	DMP	$-\overset{O}{\overset{\|}{C}}-C_{13}H_{27}$	$-\overset{O}{\overset{\|}{C}}-C_{13}H_{27}$	H
Dimyristoyl-phosphatidyl-choline	DMPC	$-\overset{O}{\overset{\|}{C}}-C_{13}H_{27}$	$-\overset{O}{\overset{\|}{C}}-C_{13}H_{27}$	$-(CH_2)_2-N^+(CH_3)_3$
Dipalmitoyl-phosphatidyl-choline	DPPC	$-\overset{O}{\overset{\|}{C}}-C_{15}H_{31}$	$-\overset{O}{\overset{\|}{C}}-C_{15}H_{31}$	$-(CH_2)_2-N^+(CH_3)_3$
Dipalmitoyl-phosphatidyl-serine	DPPSer	$-\overset{O}{\overset{\|}{C}}-C_{15}H_{31}$	$-\overset{O}{\overset{\|}{C}}-C_{15}H_{31}$	$-CH_2-CH(NH_3^+)-COO^-$

Scheme 2.2

$R^1OH + HO-P(=O)(OH)-O-P(=O)(OH)-OH \longrightarrow R^1O-P(=O)(OH)-O-P(=O)(OH)-OH \xrightarrow{\text{1) Me}_4\text{NOH}}_{\text{2) R}^2\text{Br}} R^1O-P(=O)(OH)-OR^2 \xrightarrow{\text{X-OH}}_{\text{phospholipase D}}$

X1 = (84%), (85%), (52%), (28%)

X2 = H_2N-Ser-OMe (82%), diketopiperazine (32%), H_2N-Ser-Gly-Val-OMe (31%), cytidine (69%)

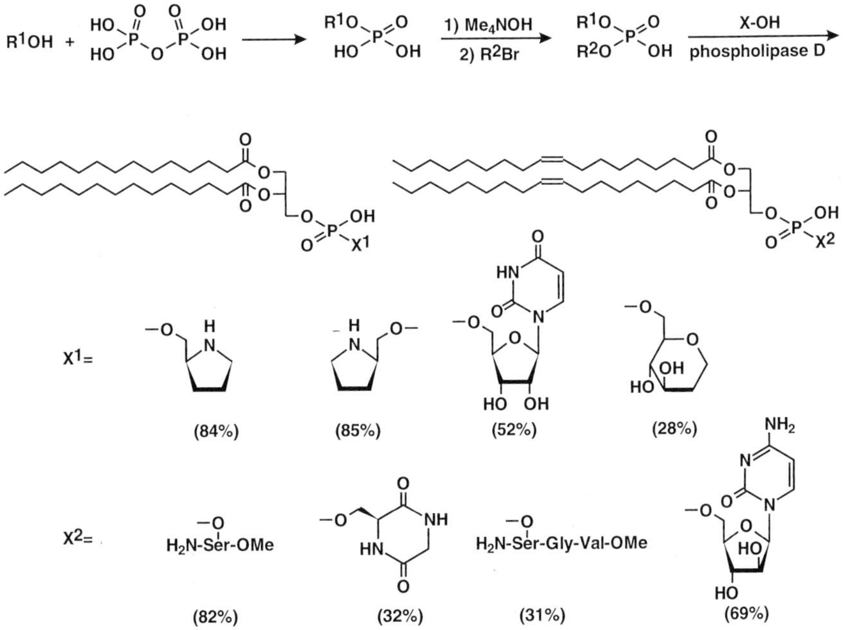

Scheme 2.3

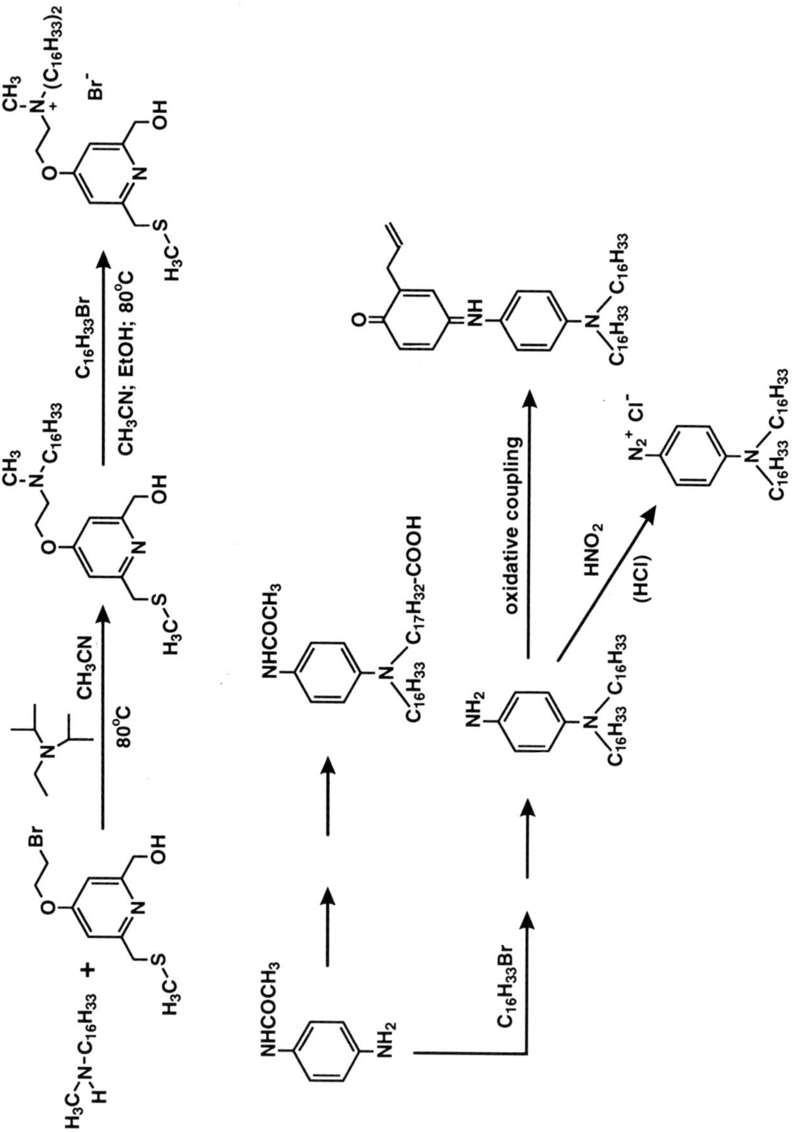

Scheme 2.4

jugates" were routinely synthesized on the gram scale by transphosphatidylation of dioleoyl- or dimyristoyl-L-phosphatidylcholine with any chosen alcohol (Scheme 2.3)[7]. Commercial phospholipase D was applied as a catalyst.

Ammonium salts with two different alkyl chains were prepared directly via subsequent alkylations of dimethylamine with primary bromides and crystallization[8]. Commercial hexadecyl-methylamine can be conveniently applied in the same way in order to convey functionality to cationic synkinons. A recent example describes subsequent alkylations with a small functional and a long-chain primary bromide (Scheme 2.4)[9]. N-acylated p-phenylenediamine was also alkylated at the second nitrogen atom which had two different alkyl chains, with or without extra functionality[10]. After deacylation, this head group can be diazotized or coupled oxidatively with various heterocycles in water (Scheme 2.4). **Photoactive and coloured membrane surfaces** are thus obtained. Phenylenediamine, pyridine and in particular N-methyl-4,4-bipyridinium chloride are relatively weak nucleophiles. Substitution of bromides is slow and the more reactive iodides can rarely be obtained commercially, but the selection of nitromethanes as solvent for bromide substitution[11] is of great help as well as the addition of sodium iodide to enforce a Finkelstein reaction[12] or a combination of both.

Primary amines react selectively with alkyl bromides[13] and acetoacetate[14] in the presence of several secondary amino groups (Scheme 2.5).

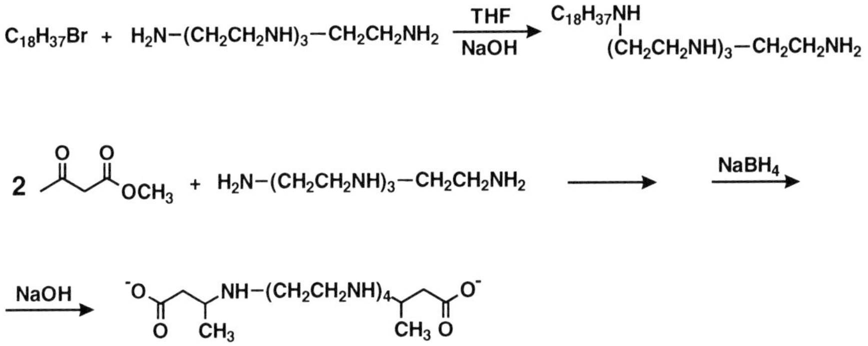

$$C_{18}H_{37}Br + H_2N-(CH_2CH_2NH)_3-CH_2CH_2NH_2 \xrightarrow[NaOH]{THF} C_{18}H_{37}\overset{|}{N}H-(CH_2CH_2NH)_3-CH_2CH_2NH_2$$

$$2 \text{ CH}_3\text{COCH}_2\text{COOCH}_3 + H_2N-(CH_2CH_2NH)_3-CH_2CH_2NH_2 \xrightarrow{NaBH_4}$$

$$\xrightarrow{NaOH} {}^-O-CO-CH(CH_3)-NH-(CH_2CH_2NH)_4-CH(CH_3)-CO-O^-$$

Scheme 2.5

Hydrosulfide end groups are usually introduced with potassium ethyl xanthogenate in acetone, thus substituting a bromide atom by a dithio carbonic acid. The latter is reductively cleaved by sodium borohydride in the presence of amines[14] (Scheme 2.6).

[7] P. Wang, M. Schuster, Y.-F. Wang, C.-H. Wong, *J. Am. Chem. Soc.*, **1993**, *115*, 10487
[8] S. Szönyi, A. Cambon, H.J. Watzke, *New J. Chem.*, **1993**, *17*, 425
[9] M.D. Everaars, A.T.M. Marcelis, E.J.R. Sudhölter, *Langmuir*, **1993**, *9*, 1986
[10] J.-H. Fuhrhop, H. Bartsch, *Liebigs Ann. Chem.*, **1983**, 803
[11] J.-H. Fuhrhop, D. Fritsch, B. Tesche, H. Schmiady, *J. Am. Chem. Soc.*, **1984**, *106*, 1998
[12] H. Finkelstein, *Chem. Ber.*, **1910**, *43*, 1528
[13] J.M. Delfino, S.L. Schreiber, F.M. Richards, *J. Am. Chem. Soc.*, **1993**, *115*, 3458
[14] S.R. Abrams, D.D. Nucciarone, W.F. Steck, *Can. J. Chem.*, **1983**, *61*, 1073

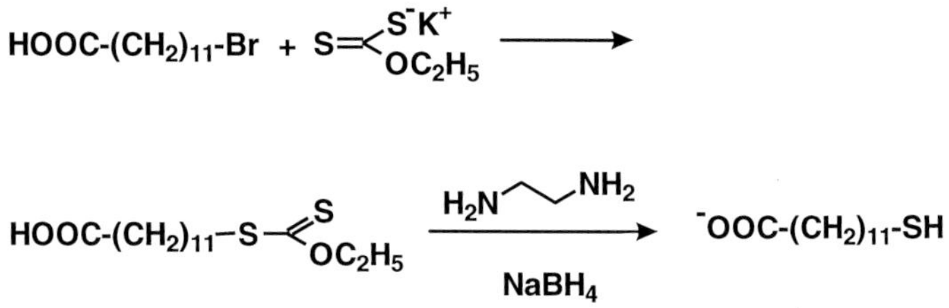

HOOC-(CH$_2$)$_{11}$-Br + S=C(S$^-$K$^+$)(OC$_2$H$_5$) ⟶

HOOC-(CH$_2$)$_{11}$-S-C(S)(OC$_2$H$_5$) $\xrightarrow[\text{NaBH}_4]{\text{H}_2\text{N} \frown \text{NH}_2}$ $^-$OOC-(CH$_2$)$_{11}$-SH

Scheme 2.6

Several of the syntheses mentioned above yielded **bolaamphiphiles** with two differing functional head groups on both ends of an alkyl chain. The so-called **"zipper" reaction** even allows the production of bolaamphiphiles from amphiphiles with a terminal methyl group. The reaction starts with a terminal alkyl to which formaldehyde is added. The resulting β-unsaturated alcohol is deprotonated by ethylenediamine/sodium hydride at the allylic CH$_2$-group next to the triple bond. A series of protonation–deprotonation reactions follows, usually leading to the ω-alkyne-α-ol in yields of 60–80%. Migration of the triple bond over 14 carbon atoms is easily achieved[14,15] (Scheme 2.7). The terminal alkyne, as well as the alcohol group, can then be employed for introducing different head groups or to elongate the hydrophobic chain.

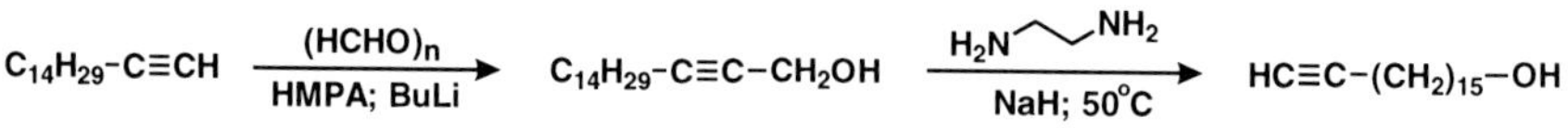

C$_{14}$H$_{29}$-C≡CH $\xrightarrow[\text{HMPA; BuLi}]{(\text{HCHO})_n}$ C$_{14}$H$_{29}$-C≡C-CH$_2$OH $\xrightarrow[\text{NaH; 50°C}]{\text{H}_2\text{N} \frown \text{NH}_2}$ HC≡C-(CH$_2$)$_{15}$-OH

Scheme 2.7

Ether linkages in open-chain **bolaamphiphiles** were obtained in a 20% isolated yield from α,ω-dibromoeicosane with an alcohol in THF containing sodium hydride[16,17] (Scheme 2.8). The synthesis of macrocyclic tetraethers was unsuccessful. Attempts to reduce macrocyclic lactones with four ester groups via a variety of methods failed[18]. The production of macrocyclic thioacetals from benzaldehyde derivatives and α,ω-dithiols[19–22] was unproblematic and produced quantitative yields (Scheme 2.8). The cyclization of 2,2-thiodiethanol with α,ω-diols in the melt and in the presence of *p*-toluenesulfonic acid is an intermediate case[18]. Apolar macrocycles were obtained in 50% yield. S-(2-hydroxyethyl)thiiranium ions are presumably formed as reactive intermediates during the ether formation steps. The sulphur atom was oxidized to the

[15] F.M. Menger, Y. Yamasaki, *J. Am. Chem. Soc.*, **1993**, *115*, 3840
[16] K. Yamauchi, Y. Sakamoto, A. Moriya, K. Yamada, T. Hosokawa, T. Higuchi, M. Kinoshita, *J. Am. Chem. Soc.*, **1990**, *112*, 3188
[17] J.-M. Kim, D.H. Thompson, *Langmuir*, **1992**, *8*, 637
[18] J.-H. Fuhrhop, U. Liman, V. Koesling, *J. Am. Chem. Soc.*, **1988**, *110*, 6840
[19] W. Autenrieth, A. Geyer, *Chem. Ber.*, **1908**, *41*, 4249
[20] C.S. Marvel, E.A. Sienick, M. Passer, C.N. Robinson, *J. Am. Chem. Soc.*, **1954**, *76*, 933
[21] C.S. Marvel, R.C. Rarrar, *J. Am. Chem. Soc.*, **1957**, *79*, 986
[22] J. Figura, Dissertationsarbeit 1992, Freie Universität Berlin

sulfoxide and sulfone. Cyclic bolaamphiphiles of perfect acid stability were thus obtained[18] (Scheme 2.8).

Amphiphiles with chiral head groups play an important role in the investigation of recognition processes in monolayers and in the construction of helical fibres. Amino acid, hydroxy acid and carbohydrate head groups were usually selected

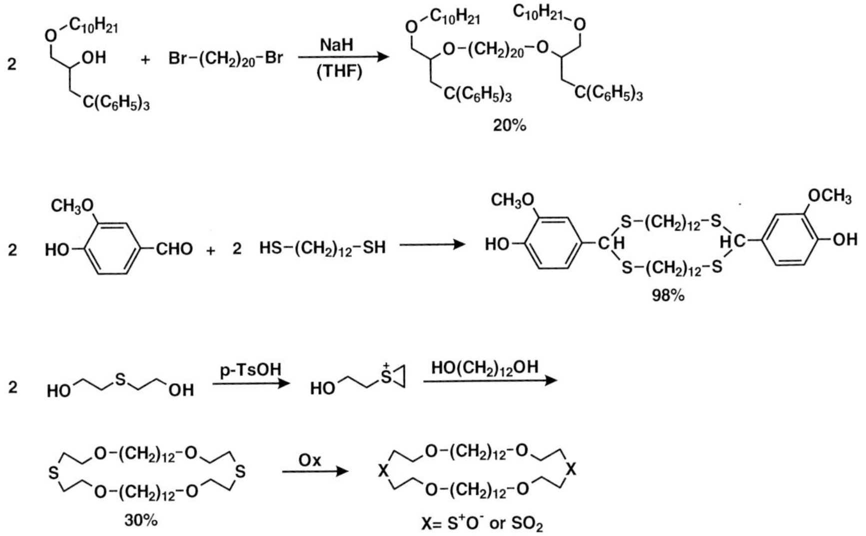

Scheme 2.8

from the chiral pool. Stearoylcysteine methyl esters and similar long-chain amino acid derivatives were obtained through the direct coupling of the acid chloride with the amino acid methyl ester[23,24] (no scheme). The synthesis of the most versatile glutamic acid triple-chain derivatives is illustrated in Scheme 2.9 together with an example of a partially fluorinated derivative[25]. L-Glutamic acid had to be di-esterified with 1H, 1H', 2H, 2H'-perfluorodecanol in the presence of p-toluenesulfonic acid in refluxing toluene before the amide group was amidated with oleoyl chloride. The resulting amphiphile with three hydrophobic chains contains two "innocent" ester groups and one strong hydrogen bonding amide group (Scheme 2.9). The final amidation reaction was also applied in the introduction of a large variety of functional head groups instead of an alkyl chain. Single-chain tartaric acid amides can be obtained in an almost quantitative yield from the diacetate of tartaric anhydride and a long-chain amine[26] (Scheme 2.9). A similar reaction with carbohydrate lactones produces open-chain glyconamides in high yields[27-29]. Michael addition of thioglycosides

[23] F.J. Zeelen, E. Havinga, *Trav. Chim. Pays-Bas*, **1958**, *77*, 267
[24] J.G. Heath, E.M. Arnett, *J. Am. Chem. Soc.*, **1992**, *114*, 4500
[25] Y. Ishikawa, H. Kuwahara, T. Kunitake, *J. Am. Chem. Soc.*, **1994**, *116*, 5579
[26] J.-H. Fuhrhop, C. Demoulin, J. Rosenberg, C. Böttcher, *J. Am. Chem. Soc.*, **1990**, *112*, 2827
[27] H.L. Frush, H.S. Isbell, *Methods Carbohydr. Chem.*, **1963**, *2*, 14
[28] W.N. Emmerling, B. Pfannemüller, *Makromol. Chem.*, **1978**, *179*, 1627
[29] J.-H. Fuhrhop, P. Schnieder, E. Boekema, W. Helfrich, *J. Am. Chem. Soc.*, **1988**, *110*, 2861

to maleic acid esters, for example the easily accessible macrolide shown in Scheme 2.9, gives amphiphiles with pyranoside head groups[30].

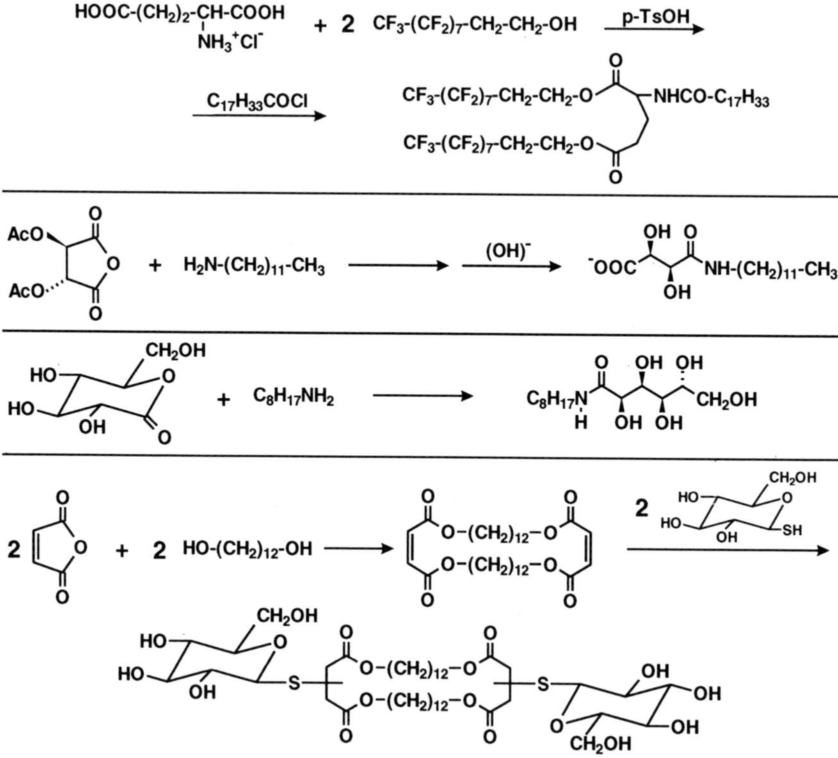

Scheme 2.9

The same maleic acid macrolide was also used to produce **bolaamphiphiles with two different head groups** in high yield. For this purpose, the macrolide was first dissolved in 2-propanol and reacted with an alkaline solution of 2-mercaptosuccinic acid. The solvent was then evaporated and the residue extracted with acetone to remove a minimal percent of the remaining macrolide. The product, which contained only one mercaptosuccinic acid substituent, was redissolved in hot 2-propanol/water 4:1 and then sodium bisulfite added. A bolaamphiphile with a "large" dicarboxylic acid head group and a "small" sulfonate head group was thus obtained in an almost quantitative yield[30,31]. Of course, it is also possible to esterify the maleic acid with bolaamphiphile alcohols in order to obtain asymmetric, non-cyclic bolaamphiphiles[32] (Scheme 2.10).

Another type of **asymmetric bolaamphiphile** not only varies the head groups but also partitions off the hydrophobic core into a hydrocarbon and a **per-**

---

[30] J.-H. Fuhrhop, H.H. David, J. Mathieu, U. Liman, H.-J. Winter, E. Boekema, *J. Am. Chem. Soc.*, **1986**, *108*, 1785
[31] J.-H. Fuhrhop, J. Mathieu, *J. Chem. Soc., Chem. Commun.*, **1983**, 144
[32] J.-H. Fuhrhop, H. Tank, *Chem. Phys. Lipids*, **1987**, *43*, 193

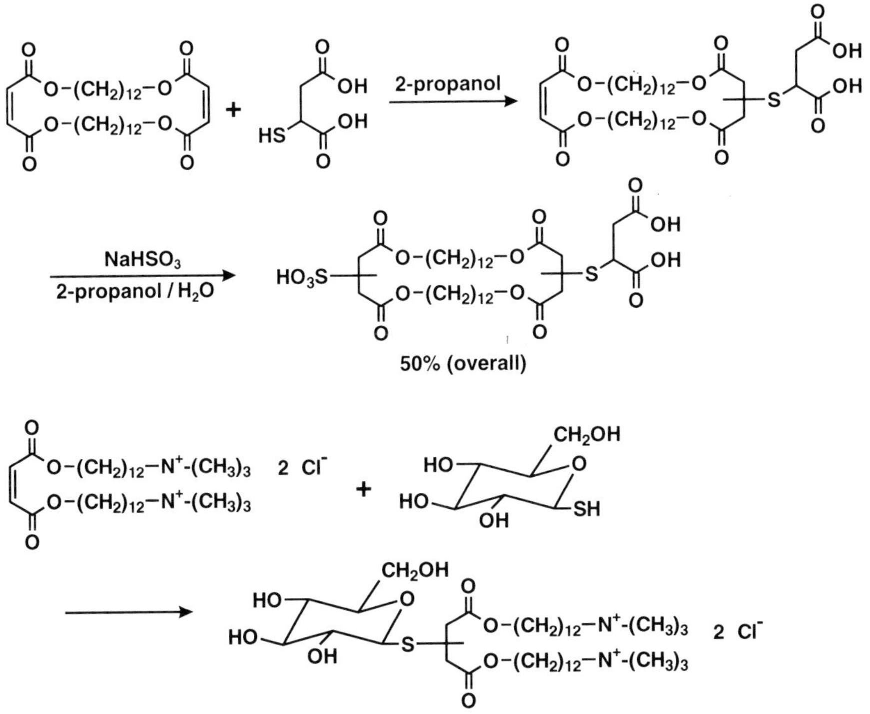

Scheme 2.10

fluorohydrocarbon part[33]. The repulsion between CF_2- and CH_2-groups then enforces the asymmetric ordering of membranes. The coupling reaction of the fluorocarbon and hydrocarbon segments involved a curious, low-yield coupling reaction of an ω-iodosulfonic acid to the double bond of 10-undecenoic acid (Scheme 2.11).

$$I\text{-}(CF_2)_8\text{-}O\text{-}(CF_2)_2\text{-}SO_3Na + CH_2=CH\text{-}(CH_2)_8\text{-}COOH$$

$$\xrightarrow[\text{(NaHCO}_3\text{ / Na}_2\text{S}_2\text{O}_4\text{)}]{CH_3CN\text{ / }H_2O} NaOOC\text{-}(CH_2)_8\text{-}CH(I)CH_2\text{-}(CF_2)_8\text{-}O\text{-}(CF_2)_2\text{-}SO_3Na$$

(low yield)

Scheme 2.11

Two amphiphiles with a reactive dye system as head groups, namely *p*-phenylenediamine and 4,4'-bipyridinium salts, have already been mentioned. Other such membrane-forming amphiphiles were modelled following natural photosynthetic membrane systems. **Carotenoid amphiphiles** are accessible from the cheap natural bolaamphiphile bixin, which carries a carboxylic acid on one end and a methyl ester on the other. A variety of orange-coloured bolaamphi-

---

[33] K. Liang, Y. Hui, *J. Am. Chem. Soc.*, **1992**, *114*, 6588

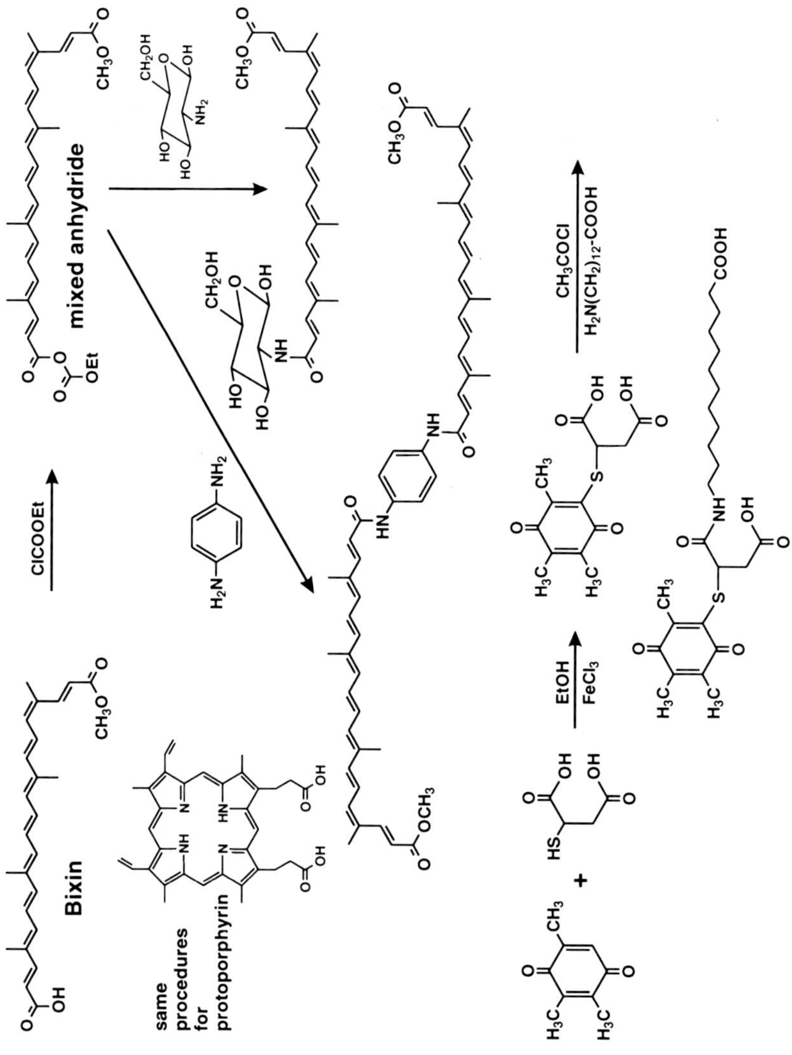

Scheme 2.12

philes were synthesized by formation of the mixed anhydride of bixin with ethyl chloroformate and subsequent amidation[34] (Scheme 2.12). **Quinone** bola-amphiphiles were either obtained via Michael addition of thiols to quinones and subsequent oxidation, or by condensation reactions of aminoquinones with mixed carboxyanhydrides[35] (Scheme 2.12). Porphyrin amphiphiles are easily accessible through amidation of mixed anhydrides, usually made up of proto-**porphyrin** derivatives and ethyl chloroformate[36]. In all these cases, it is advisable to avoid activation with acid chlorides as the dyes usually do not survive.

Surprisingly simple Friedel–Crafts reactions introduce **ferrocene** units into **amphiphiles**. For example, 11-bromo-undecanoyl chloride can be added directly to ferrocene in dichloromethane with aluminium chloride as a catalyst. The corresponding (11-bromo-undecanoyl) ferrocene can be isolated via simple recrystallization, and the ketone removed by amalgamated zinc (does not remove the bromide). The bromide can finally be substituted by alcohols and amines to yield ethers and ammonium salts[37] (Scheme 2.13).

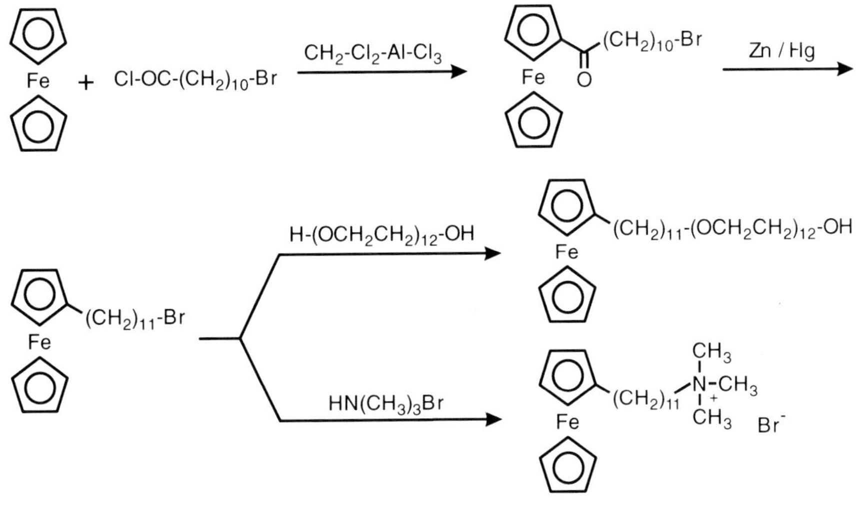

Scheme 2.13

The final class of important synthetic lipids are those with a **polymerizable** substituent. A significant concept present is the introduction of a **hydrophilic spacer group** between the hydrophobic core and the polymerizable group. Only in polymers made of "spacer lipids" does the polymer chain remain flexible enough to decouple its movements from those of the membrane in order to

---

[34] J.-H. Fuhrhop, M. Krull, A. Schulz, D. Möbius, *Langmuir*, **1990**, *6*, 497
[35] J.-H. Fuhrhop, H. Hungerbühler, U. Siggel, *Langmuir*, **1990**, *6*, 1295
[36] J.-H. Fuhrhop, C. Demoulin, C. Böttcher, J. Köning, U. Siggel, *J. Am. Chem. Soc.*, **1992**, *114*, 4159
[37] T. Saji, K. Hohsino, Y. Ishii, M. Goto, *J. Am. Chem. Soc.*, **1991**, *113*, 450

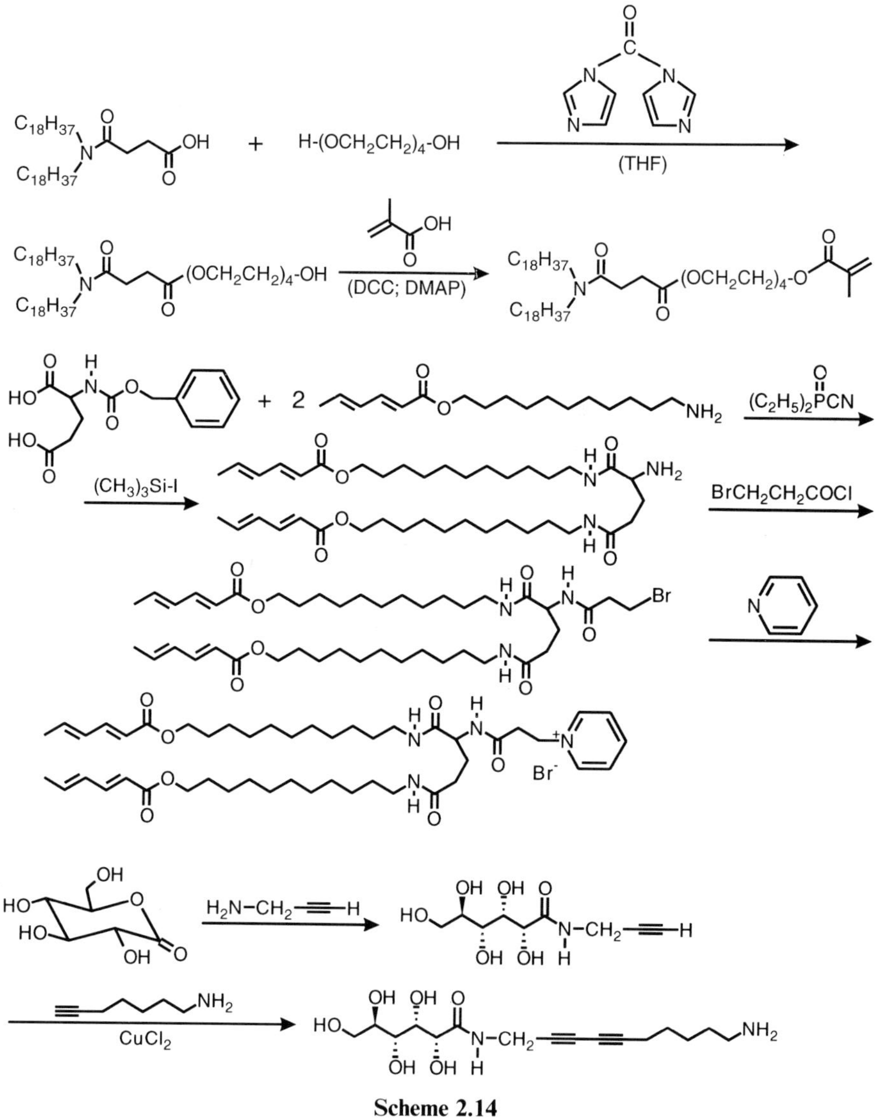

Scheme 2.14

build polymeric LB-multilayers and vesicles[38]. The synthesis of the double chain amine given above implied direct condensation of N,N-dioctadecylsuccinamide with tetraethylene glycol and carbonyldiimidazole as a catalyst. Esterification of the remaining primary alcohol group with methacrylic acid was carried out with dicyclohexylcarbodiimide (DCC) and 4-(dimethylamino)-pyridine (DMAP)[38] (Scheme 2.14). Spacers, however, are not required in fibres. Micellar fibres of extremely high molecular weight were obtained from double-

---

[38] R. Elbert, A. Laschewsky, H. Ringsdorf, *J. Am. Chem. Soc.*, **1985**, *107*, 4134

chain glutamic acid dienes[39] and gluconic acid diynes[40] (Scheme 2.14). Polymerization always transpired via UV irradiation of preformed monolayers, vesicles or fibres.

Reversible polymerization or cross-linking by oxidative coupling of hydrosulfides to sulfides is often employed in biological protein chemistry. This principle can be also realized in membrane assemblies, where two terminal SH-groups are introduced into double-chain amphiphiles[41] (Scheme 2.15). Vesicles made from such amphiphiles form (on average) 20-mers with hydrogen peroxide and are depolymerized by 1-octanethiol.

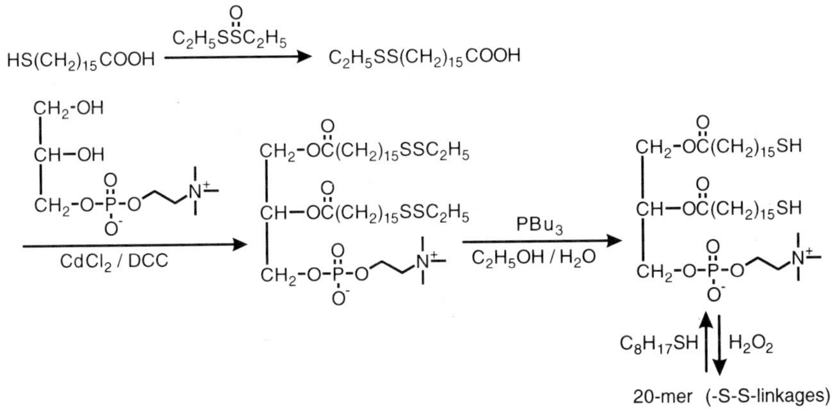

Scheme 2.15

---

[39] H. Ihara, M. Takafuji, C. Hirayama, D.F. O'Brien, *Langmuir*, **1992**, *8*, 1548
[40] J.-H. Fuhrhop, P. Blumtritt, C. Lehmann, P. Luger, *J. Am. Chem. Soc.*, **1991**, *113*, 7437
[41] N.K.P. Samuel, M. Singh, K. Yamaguchi, S.L. Regen, *J. Am. Chem. Soc.*, **1985**, *107*, 42

CHAPTER 3

Micelles and Microemulsions

3.1 Introduction

Names given to differentiate between supramolecular mono- and bilayer assemblies should be clear and relate to structure in a simple and general way. Micelle (Latin *mic(a)*, grain; + *ella*, diminutive suffix) indicates nothing but the ultimate smallness of these assemblies. The word microemulsion (Greek *micro*, small; Latin *emulsus*, milked out) evokes a milk-like aqueous system containing (fat) droplets similar to those found in milk but much smaller. Both names produce, to a first approximation, an appropriate impression of the real systems.

On the supramolecular level, micelles and microemulsions are, structurally, of no use to molecular architecture. Their short-lived structures are only vaguely determined by non-directional repulsive forces and by solvophobic effects. Organic chemists should therefore be advised to keep their distance from such systems and leave them to physical chemists, who describe them perfectly through thermodynamic, kinetic and mechanical laws. Unfortunately, though, physical chemists could not resist the temptation to establish structural models of micelles, so on this point, the supramolecular chemist should express his disapproval! Furthermore, preparation procedures of better defined supramolecular systems often begin with micelles, which also appear as side- or decomposition products. Finally and most importantly, although micelles and microemulsions are not well-defined structures on the molecular level, they are often most useful in the solubilization of hydrophobic dyes and reductants in water. As long as these systems are of interest to organic chemists we will continue to describe them, but refer the more physically oriented reader to other books[1–5].

---

[1] P.H. Elworthy, A.T. Florence, C.B. MacFarlane, Solubilization by Surface Active Agents, Chapman and Hall, London, **1968**
[2] J.H. Fendler, E.J. Fendler, Catalysis in Micellar and Macromolecular Systems, Academic Press, New York, **1975**
[3] R.D. Vold, M.J. Vold, Colloid and Interface Chemistry, Addison–Wesley, London, **1983**
[4] D.M. Small, The Physical Chemistry of Lipids, Plenum Press, New York, **1986**
[5] J.H. Clint, Surfactant Aggregation, Blackie, Glasgow, **1992**

3.2 Why Micelles are Formed and Remain in Solution

Micelles are loose aggregates of amphiphiles in water or organic solvents which form above a certain temperature (Krafft point) and concentration (critical micellar concentration, cmc). Below the Krafft temperature, clear micellar solutions become turbid and the amphiphile forms three-dimensional hydrated crystals. Below the cmc, micelles dissociate into monomers and small aggregates. Above the cmc, the micelles of an aggregation number n are formed: n then remains stable over a wide concentration range[1-5]. Table 1 gives some typical cmcs and three Krafft point values.

Micellar solutions are stable and remain clear over years, although the individual micelle usually explodes and reforms within a few milliseconds[2]. The question arises as to why such aggregates of a limited number of molecules (usually between fifty and one hundred) form at all?

The first reason lies in the fact that the interaction between solvent molecules (usually water) is stronger than the interaction between the solvent and the solute[6]. This effect alone would lead to a precipitation of the solute. In the case of amphiphiles which form micelles, however, the head groups are strongly hydrated and repulse each other. The hydration forces and steric forces[7] which are made responsible for this repulsion effect prevent crystallization above the Krafft point and also above the cmc. Where the formation of 3D crystals is impeded, the smallest possible droplet is formed, removing the alkyl chains from the solvent. The interactions between solvent molecules are therefore disturbed to a minimal extent, allowing the head groups to be solvated with a minimal entropy loss. It is irrelevant whether the solvent contains clusters or not. Micelle formation only occurs as a result of a solvation of head groups and non-solvation of a solvophobic core[6].

There is no simple theory of solute–solvent interaction. Experimentally, however, it has been shown that the insolubility of hydrocarbons in water is indeed due to entropy effects. For example, ΔG for the transfer of bulk butane into water splits up into $\Delta H = -4.2$ kJ/mol and $-T\Delta S = +28.7$ kJ/mol. The decrease in entropy, caused by a more ordered structure of the surrounding water molecules, thus contributes 85% to this interaction[7]. Computer simulations indicated that water molecules in nonpolar regions have the same number of strong hydrogen bonds as those in the "bulk". These bonds are, however, only distributed in an average sense among 4.7 neighbours within a 3.5 Å distance, whereas in bulk water 5.75 neighbours reside within that distance. Thus, relative to the bulk, a more developed bonding network exists among the water molecules in the immediate vicinity of a nonpolar group. **The low solubility of nonpolar molecules and the entropic character of ΔG is known as hydrophobic effect** (Figure 3.1c). It leads to a much stronger interaction energy between hydrophobic molecules in water than the van der Waals interaction in free space[7]. The favoured orientation of water at the surface of a nonpolar

---

[6] Y. Ishikawa, H. Kuwahara, T. Kunitake, *J. Am. Chem. Soc.*, **1994**, *116*, 5579
[7] J.N. Israelachvili, Intermolecular and Surface Forces, Academic Press, London, **1985**

Table 1 CMCs and Krafft points of several surfactants

Compound	cmc (mmol/L)	n	Krafft point (°C)
$C_{12}H_{25}SO_4Na$ (SDS or NaLS)	8.1	62	9
$(C_{12}H_{25}SO_4)_2Ca$	—	—	50
$C_{12}H_{25}N(CH_3)_3Br$	14.4	50	—
$C_{12}H_{25}COOK$	12.5	50	—
$C_{12}H_{25}(OCH_2CH_2)_6OH$	0.087(!)	400(!)	—
$C_{16}H_{33}N(CH_3)_3Br$ (CTAB)	0.92	23	61

solute has also been calculated. If θ is the angle between the O—H bond or O-lone pair direction in water and the axis defined by the methyl carbon and water oxygen, the ideal orientation[8] corresponds to θ = 0° (Figure 3.1a). This molecular orientation is typical for crystalline clathrate hydrate compounds such as CH_4–H_2O crystals[9]. In other solvents θ is generally defined as the angle between any (partial) charge of the solvent molecule, the solvent-centre of mass, and the solvophobic group.

For the solvent molecule near a solvophilic group, the total number of neighbours is not reduced, because the solute can take the place of a solvent molecule. Model calculations of the solvation energies of water-dissolved amino acid and nucleic acid base molecules, for example, yielded contour line diagrams, in which the distance of the first water layer (binding energy = 0) to polar and nonpolar groups is about the same. In the environment of polar groups, however, the water molecules are much more strongly bound as indicated by many contour lines which were drawn for binding energies of 1, 2, 3, 4 kcal/mol. These contour line diagrams generally indicate that carboxyl groups and polar heterocycles are strongly hydrated, where NH_2 and OH groups are not[10] (Figure 3.1b). Dehydration and particle growth (Figure 3.1c) is therefore strongly chemoselective.

The appearance of hydrogen bonded clusters in water (Figure 3.2) plays an important role in the solubilization of charged molecules (= electrolytes). Sodium and chloride ions can, for example, only be efficiently separated in water, as this solvent has an enormously high dielectric constant of 81. Ions are not only solvated by a few single solvent molecules, as in organic solvents, but by whole clusters. As a consequence, micellar structures of electroneutral amphiphiles can also successfully exist in organic solvents, but charged surfaces call for hydration forces, clusters and a large dielectric constant. The first water layer around a hydrated ion is usually organized in a tetrahedral or octahedral fashion. Small ions (Li^+, Na^+, Mg^{2+}, Ca^{2+}) bind a second layer of water molecules in crystalline order, but not the larger ions (K^+, Cl^-).

[8] P.J. Rossky, M. Karplus, *J. Am. Chem. Soc.*, **1979**, *101*, 1913
[9] F. Franks, P.S. Reid, in F. Franks (ed), Water, Vol. 2, Plenum, New York, **1975**
[10] E. Clementi, F. Cavallone, R. Scordamanglia, *J. Am. Chem. Soc.*, **1977**, *99*, 5531, 5545

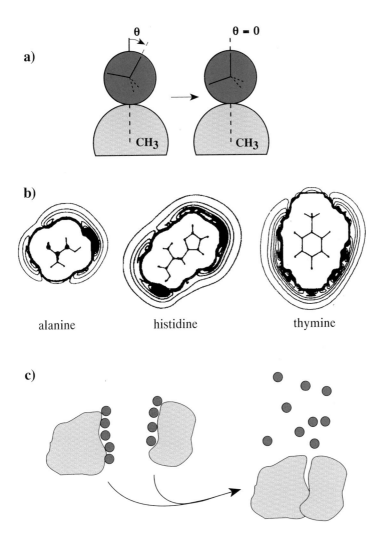

Figure 3.1 *Schematic representations of a) a water molecule orientation near a nonpolar CH_3-group[9], which is optimal if none of the hydrogen atoms or electron pairs is directed toward the nonpolar group ($\theta = 0$); b) contour line diagrams of three polar molecules with the first inner line of a solvation energy of 0 kcal/mol, the second line of 1 kcal, the third line of 2 kcal/mol etc[10]; and c) of the hydrophobic effect. Upon association of hydrophobic particles water or other solvent molecules are released. Entropy grows.*

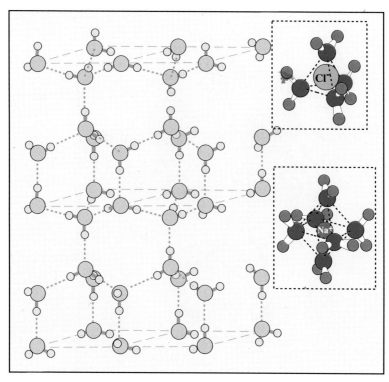

Figure 3.2 *An idealized, ice-like water cluster and the first hydration spheres of Na^+ and Cl^- ions.*

3.3 Structural Models of Micelles and the Wetness Problem

Sedimentation, diffusion and light scattering data, as well as well-fixed aggregation numbers, point to a finite geometrical shape of micelles: spheres or cubes are most likely[11]. Laser Raman spectra of aqueous NaLS, CAD and CTAB* solutions reflect a fluid-like core, with a predominance of *gauche* rotamers[12]. The outer CH_2 groups, however, occur predominantly in an *all-trans* conformation[12]. ^{13}C-NMR T_1 values of carbon atoms also point to motions of the inner segments approaching those of the liquid hydrocarbons[13].

These data tend to suggest a statistical lattice model[14,15] (Figure 3.3). Extremely high conformational energy is however implied here, since, in contrast to the NMR results, most of the outer segments occur in energy-rich *gauche* conformations. Furthermore, the experimental aggregation numbers

---

* NaLS = SDS = sodium lauryl/dodecyl sulfate; CAD = DAC = dodecylammonium chloride; CTAB/CTAC = cetyltrimethylammonium bromide/chloride
[11] F.M. Menger, *Acc. Chem. Res.*, **1979**, *12*, 111
[12] K. Kalyanasundaram, J.K. Thomas, *J. Phys. Chem.*, **1976**, *80*, 1462
[13] E. Williams, B. Sears, A. Allerhand, E.H. Cordes, *J. Am. Chem. Soc.*, **1973**, *95*, 4871
[14] K.A. Dill, P.J. Flory, *Proc. Natl. Acad. Sci. USA*, **1981**, *78*, 676
[15] D.W.R. Gruen, *J. Colloid Interface Sci.*, **1981**, *84*, 281

Micelles and Microemulsions

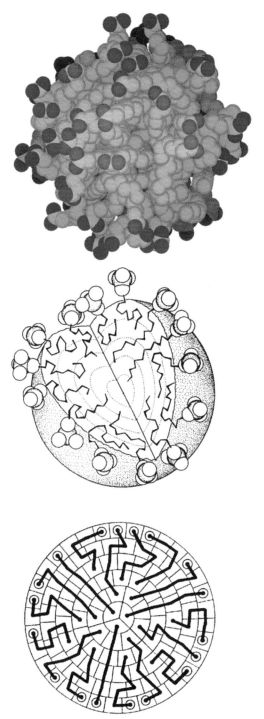

Figure 3.3 *3D and cross-sectional models of SDS micelles with a statistical conformation of hydrocarbon chains of 60 molecules*[14,15]. *(SDS = sodium dodecyl sulfate)*.

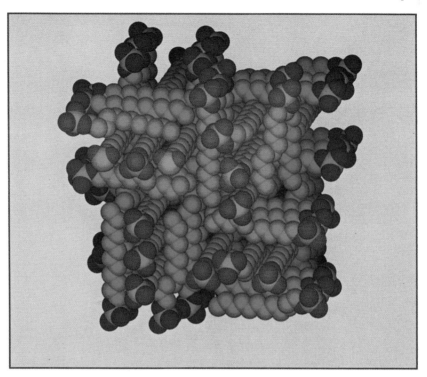

Figure 3.4 *3D block model with essentially stretched conformations of 72 molecules[16].*

are much larger than those of the model. Another model arranges the parallelly ordered surfactant molecules into blocks[16] which lie perpendicularly to each other. The width of each block equals the length of the oligomethylene chains (Figure 3.4). The hydrocarbon chain length measured in units of hydrocarbon chain diameter governs the aggregation number. Both experiment and the block model agree in this respect. The block model, however, fits so tightly that the ready intrusion of large dye molecules (typical for micellar solutions) cannot be understood whereas a model in which each molecule behaves independently is much more likely.

In the most realistic model of the micelle, 80% of its volume is composed of a reef structure and in the remaining 20% which is present in the centre, the hydrocarbon chains really come into close contact and cement the aggregate together[11]. In the diffuse "pincushion" configuration (Figure 3.5), the aggregation number is less determined by geometry than by repulsion between the hydrated head groups. Dissolution of hydrophobic molecules is, however, favoured by the displacement of water molecules from the inner parts of the micelle, thus evoking a tendency to avoid the hydrophobic gaps.

All three models imply that the majority of the hydrophobic part of the amphiphiles comes in contact with water or is "wet". This "wetness" is particularly evident even in the terminal carbon atoms of the alkyl chains

[16] P. Fromherz, *Ber. Bunsenges. Phys. Chem.*, **1981**, *85*, 891

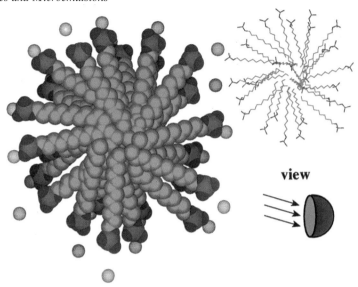

Figure 3.5 *Cross sections of the "reef" or "rugged" micelle model with 60 molecules in essentially stretched conformations*[11].

resulting from NMR experiments with **1–3**[17,18]. It is unlikely that either the acetylenic or the benzoic groups should disturb the structure of micelles in water and normal cmc-values have indeed been found. Nevertheless, NMR spectra revealed chemical shifts of the acetylenic protons in D_2O at $\delta = 2.1$–2.2 ppm, typical for protic solvents and quite a distance downfield from the 1.6–1.8 ppm in hydrocarbon solvents. Furthermore, the pK_a of the benzoic acid in bolaamphiphile **3** is 3.73 in water at concentrations where the compound is totally monomeric. In the vicinity of the cmc, the pK_a remains 3.78, indicating no trace whatsoever of hydrophobic protection of the benzoic acid moiety, meaning that water is present throughout in micelles, even in the inner core.

$$HC\equiv CCD_2\text{-}(CH_2)_{10}\text{-}N^+(CH_3)_3$$
1

$$HC\equiv CCD_2\text{-}(CH_2)_{10}\text{-}OSO_3^-$$
2

$$Cl^-Me_3N^+(CH_2)_{16}O\text{—}\underset{COOH}{\text{C}_6H_3}\text{—}O(CH_2)_{16}NMe_3^+\,Cl^-$$
3

High resolution neutron scattering on SDS micelles in water did not contribute towards the clarification of the picture. A 5 Å resolution study, however,

---

[17] F.M. Menger, J.F. Chow, *J. Am. Chem. Soc.*, **1983**, *105*, 5501
[18] F.M. Menger, C.E. Mounier, *J. Am. Chem. Soc.*, **1993**, *115*, 12222

showed considerable fluctuations from the average structure and deuterated methyl groups at the end of the dodecyl chains were shown *not* to be concentrated near the centre of the micelle[19].

3.4 No Known Surfactant Has a Conical Shape

Structural modelling can be a very formal affair! Take, for example, an ideal sphere with a bimolecular diameter in the shape of a micelle and divide it by the aggregation number. A cone (Figure 3.6) with a surface a_o of the circle on the top, a length l_c and a volume V is obtained. So far, so good. But now, to make sense out of the cone, a_o is called the head group area, l_c a critical length "specified for a given lipid", and v becomes the hydrocarbon chain volume. As a result, one now has defined nicely an "average molecule", using the shape of an aggregate which is less than ill-defined (see Figure 3.3–3.5). This does not sound very promising and true enough, things get worse! A "critical packing parameter" is defined via the cone's measurements and conclusions such as "SDS in low salt is a cone" and "SDS in high salt is a truncated cone" are drawn[7]. Such pseudo-structural, mean-dynamic-packing models do not rationalize the appearance of molecular assemblies, but mystify them. Not a single crystal structure, NMR spectrum or molecular model gives evidence of a cone-like shape of any known amphiphile and one cannot derive it from a molecular assembly, for which the models are as far apart as spherical droplets, cubic blocks and irregular reefs.

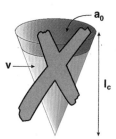

Figure 3.6 *Amphiphilic molecules with the "dynamic shape" of a cone are often evoked in literature[7] to explain the formation of micelles and micellar fibres. No such molecules exist, neither as individual molecules nor as an average of existing conformers.*

Later on we shall portray several examples where very similar amphiphiles, even diastereomers and enantiomers, produce a variety of molecular assemblies under identical conditions. It is not the shape of a molecule that determines the shape of its molecular assemblies, but the degree of binding and repulsion between them. The more binding interactions occurring between molecules, the larger will be the assemblies formed, because monomers are less likely to

[19] B. Cabane, R. Duplessix, T. Zemb, *J. Physique*, **1985**, *46*, 2161

escape. Curvature of large assemblies, such as micellar fibres, is then determined by the degree of bending within the monomers (see page 123).

3.5 The Millisecond Life of Individual Micelles and their Featureless Appearance

Micelles are a dynamic species, which rapidly break up and reform. In a microsecond time scale, single surfactant molecules retreat from and travel towards micelles, and within milliseconds the whole micelle disintegrates and reassembles[20].

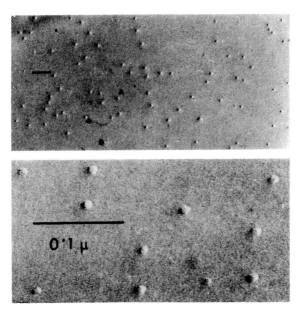

Figure 3.7 *Transmission electron micrographs of eicosane sulfate micelles in vitreous ice (two different magnifications)*[22].

Attempts to characterize micelles by electron microscopy often fail. 5 nm lysolecithin micelles, however, can be fixated by negative staining with tungstate[21]. Eicosane sulfate gave 8–12 nm micelles upon rapid cryo-fixation and freeze etching (Figure 3.7)[22] and a nonionic surfactant of high molecular weight (HCO-60) produced 10 nm globules in vitreous ice[23]. Similar micelles were also found in aqueous solutions of a long chain α,ω-bipyridinium tetrabromide bolaamphiphile when it was precipitated from aqueous solution as a tetraperchlorate[24]. Monolayer platelets were accompanied by a large number of spherical particles with diameters close to 10 nm. CTAB micelles were also

---

[20] J.H. Fendler, Membrane Mimetic Chemistry, Wiley, New York, **1982**, p. 25f
[21] K. Inoue, K. Suzzki, S. Nojima, *J. Biochem.*, **1977**, *81*, 1097
[22] L. Bachmann, W. Dasch, P. Kutter, *Ber. Bunsenges, Phys. Chem.*, **1981**, *85*, 883
[23] J.L. Burns, Y. Cohen, Y. Talmon, *J. Phys. Chem.*, **1990**, *94*, 5308
[24] J.-H. Fuhrhop, D. Fritsch, B. Tesche, H. Schmiady, *J. Am. Chem. Soc.*, **1984**, *106*, 1998

directly imaged by cryomicroscopy in vitrified films[25]. Resolution was poor. More interestingly, worm-like micelles without endings were spotted in 0.37 wt % CTAB solutions containing 3.0 wt % of NaBr electrolyte (see section 5.3). Micelles formed by 11-ferrocenylundecyl polyoxyethylene ether were positively stained with osmium tetroxide[26]. The PEG-part then turned black and the ferrocene–undecane part turned out as white featureless spots. The result is again similar to negative staining as no defined osmium ring was detectable.

Micelles can therefore form an image under the electron microscope, but with the exception of approximate radii and spherical shapes, nothing else could be seen so far. Reasonable STM or AFM images have, to the best of our knowledge, not yet been obtained.

3.6 The Dissolving of Large Molecules via Micelles – One or Two per Micelle

One particular property of micelles stands out above all others: their ability to solubilize organic compounds in water. Benzene, for example, dissolves in SDS to the extent of 0.90 mol/mol surfactant, resulting in around 40 benzene molecules per micelle[11]. NMR chemical shift data situate most of the benzene at the micelle–water interface[27], but the localization of small solubilizates in micelles is never uniform.

Molecules, both large and apolar can, however, be isolated in micelles. Saturation concentrations for porphyrins, pyrene and similar hydrophobic dyes in SDS are of the order of 1×10^{-4} mol/dm^3. Statistically, this can be seen to be about 0.5–1.0 dye molecules per micelle. A porphyrin with four dodecyl sulfate side-chains dissolves as a micellar cluster in water and produces a single broad ESR peak for the central copper(II) ion[28] (Figure 3.8). In the presence of SDS micelles the nine line ESR signal of the monomer can be seen. Metal-free porphyrin bases and their main group metal complexes strongly fluoresce in micellar solutions. This also indicates a low degree of aggregation (compare section 5.8).

Porphyrins dissolved in SDS micelles have also been applied towards the determination of pK_a values[29] of the central atoms, which are around 5.5 for alkyl substituted porphyrins, but drop to 4.8 for divinyl substituted protoporphyrin dimethyl ester and to 3.0 for porphyrins with two β-pyrrolic carbethoxy or formyl substituents.

SDS micelles with dissolved thyminyloctyl ω-ammonium salts form 1:1 complexes with N-6-acetyl-9-propyl adenines (Figure 3.9). The synkinesis of

---

[25] P.K. Vinson, J.R. Bellare, H.T. Davis, W.G. Miller, L.E. Scriven, *J. Colloid Interface Sci.*, **1991**, *142*, 74
[26] S. Yokoyama, H. Kurata, Y. Harima, K. Yamashita, K. Hoshino, H. Kokado, *Chem. Lett.*, **1990**, 343
[27] J.C. Eriksson, G. Gilberg, *Acta Chem. Scand.*, **1966**, *20*, 2019
[28] J.-H. Fuhrhop, M. Baccouche, *Liebigs Ann. Chem.*, **1976**, 2058
[29] W.S. Caughey, W.Y. Fujimoto, B.P. Johnson, *Biochemistry*, **1966**, *5*, 3830

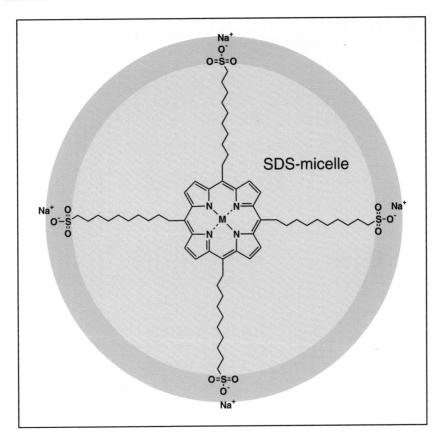

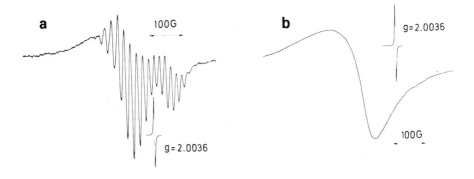

Figure 3.8 *An SDS-dissolved copper porphyrin gives the ESR spectrum a) of a monomer, whereas the same amphiphilic porphyrin in water produces b) the spectrum of a micellar aggregate*[28].

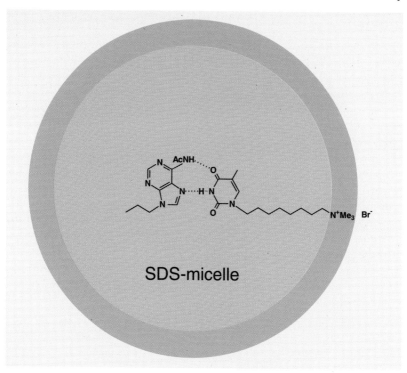

Figure 3.9 *Hydrophobic A–T derivatives form a heterogeneous dimer in SDS micelles*[30].

this system is presumably successful, because only one long-chain thymine is bound per micelle whereas the short-chain adenine is then fixated by the thymine base. Base stacking, which occurs in aqueous solution, is not possible as the micellar volume is too small. The ^{1}H-NMR chemical, downfield shift of the thymine imino proton upon hydrogen bonding has been used as a sensitive probe of its environment[30]. A tailored excitation pulse sequence was necessary in order to suppress the water peak. The measured association constant of the hydrophobized A–T pair in micelles was 16 M^{-1}.

SDS micelles were also employed for the light-induced cyclization of polyenes to steroids in low yield[31]. No such reaction was observed in homogeneous organic solutions. Again, it is believed that only one polyene molecule is dissolved within one micelle. On the other hand, *trans*-cinnamic acid produced dimeric products after UV irradiation in 1% aqueous CTAB, while no photodimers were formed in homogeneous solutions[32].

[30] J.S. Nowick, J.S. Chen, G. Noronha, *J. Am. Chem. Soc.*, **1993**, *115*, 7636
[31] U. Hoffmann, Y. Gao, B. Pandey, S. Klinge, K.-D. Warzecha, C. Krüger, H.D. Roth, M. Demuth, *J. Am. Chem. Soc.*, **1993**, *115*, 10358
[32] Y. Nakamura, *J. Chem. Soc., Chem. Commun.*, **1988**, 477

3.7 Light-Initiated Electron Transfer in Micelles

The basis for numerous attempts to degrade water or to produce an electric current by light is the photoproduction and quick separation of reactive ions in micellar systems. The first results showed that aqueous micellar systems strongly promote the photoionization of pyrene[33], phenothiazine[34], and tetramethyl benzidine[35]; the energy for photoionization being low in this case, as the micellar medium is polarized by the dye cation formed and because the electron may escape into the aqueous medium[36]. Negative surfaces of anionic micellar charges inhibit the back-reaction neutralization and effective charge separation is the outcome. An example can be seen in the photoinduced reduction of cupric ions in water by N-methylphenothiazine ($=$ MPT) in SDS micelles[37]. The transfer of an electron from MPT to Cu^{2+} occurs in under 1 nanosecond. The cuprous ion formed in the photoredox process escapes from the micelle into the bulk solution before back transfer of electrons to MPT^+ can occur. A second redox reaction with the $[Fe(CN)_6]^{3-}$ anion destroys the Cu^+ ions. The oxidized MPT^+ in the micelle remains stable due to the ferrocyanide being kept at a distance from the anionic micellar surface (Figure 3.10). The light energy is then stored in the redox couple $MPT^+ [Fe(CN)_6]^{4-}$.

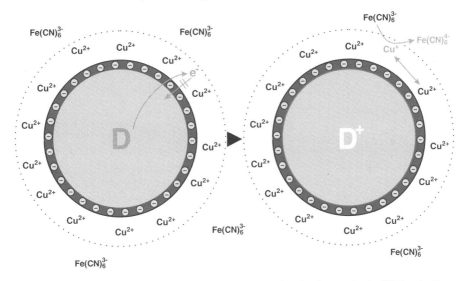

Figure 3.10 *A dye D ($=$ N-methylphenothiazine) dissolved in anionic SDS micelles donates electrons to Cu^{2+} counterions upon UV irradiation. The back-reaction is slowed down by the micelle's surface charge and becomes essentially irreversible when the copper(I) ion looses its electron to ferricyanide anions*[37].

---

[33] M. Grätzel, J.K. Thomas, *J. Phys. Chem.*, **1974**, *78*, 2248
[34] S.A. Alkaitis, G. Beck, M. Grätzel, *J. Am. Chem. Soc.*, **1975**, *97*, 5723
[35] S.A. Alkaitis, M. Grätzel, *J. Am. Chem. Soc.*, **1976**, *98*, 3549
[36] J.K. Thomas, P. Piciulo, *J. Am. Chem. Soc.*, **1978**, *100*, 3239
[37] Y. Moroi, A.M. Braun, M. Grätzel, *J. Am. Chem. Soc.*, **1979**, *101*, 567, 573

Cationic redox pairs were also stabilized in micellar systems[38]. Photoexcited ruthenium(II) dipyridinate acts as a strong, one-electron reductant and reduces 4-methyl-4'-tetradecyl-bipyridinium ($C_{14}MV^{2+}$) dichloride to the corresponding radical ($C_{14}MV^{+}$). The back reaction between Ru^{III} and $C_{14}MV^{+}$ can be slowed down by the addition of CTAC micelles, which entrap $C_{14}MV^{+}$ much more efficiently than the dication $C_{14}MV^{2+}$ and which also repel Ru^{3+} ions.

3.8 Micelles Made of Rigid Amphiphiles

A number of amphiphiles with a rigid hydrophobic core also form micelles. A cationic anthraquinone dissolved in aqueous 0.2 mol/L LiOH gives micelles with an apparent cmc of 7.4×10^{-5} mol/L. Although the position of the UV band at 330 nm hardly changes upon micellization, the modification in the apparent molar extinction coefficient at the cmc is drastic[39]. The same copper porphyrin which dissolved extremely well in SDS micelles (see Figure 3.8) forms micelles itself in distilled water and then copper–copper interactions produce a very broad ESR signal (Figure 3.8)[38]. The **"facial amphiphile" 4** carrying the head group not at the end, but in the centre of a hydrocarbon core, also forms micelles in water. It dissolves orange-coloured OT at a critical concentration of 6×10^{-2} M. **5** without the short side-chain has no such effect[40].

The above mentioned examples indicate that a flexible hydrocarbon chain is *not* a prerequisite for micellization.

The only other well-known micelles possessing a rigid hydrophobic core are those based on **bile acids, some of the most important biological detergents**. Particularly efficient solubilizing agents of the bile acid family are the salts of deoxycholic acid **6**[41–44]. Their complexes with water-insoluble compounds are called "choleic acids". The aggregation number of deoxycholic acid (DCA) in distilled water is approximately 10–12 ("primary micelles") and rises to about 100 ("secondary micelles") in more concentrated solutions and/or after addition of electrolytes to primary micelles. The cmc of primary micelles[43] lies in the range of $1–5 \times 10^{-3}$ mol/L. Less polar cholic acids are in general much better

---

[38] P.-A. Brugger, M. Grätzel, *J. Am. Chem. Soc.*, **1980**, *102*, 2461
[39] K. Hoshino, T. Saji, K. Suka, M. Fujihara, *J. Chem. Soc., Faraday Trans. I*, **1988**, *84*, 2667
[40] D.G. Barrelt, S.G. Gellman, *J. Am. Chem. Soc.*, **1993**, *115*, 9243
[41] H. Wieland, G. Sorge, *Hoppe–Seyler's Z. Physiol. Chem.*, **1916**, *97*, 1
[42] J.C. Carey, J.C. Montet, M.C. Phillips, M.J. Armstrong, N.A. Mazer, *Biochemistry*, **1981**, *20*, 3637
[43] A. Roda, A.F. Hofman, K.J. Mysels, *J. Biol. Chem.*, **1983**, *258*, 6362
[44] D. Leibfritz, J.D. Roberts, *J. Am. Chem. Soc.*, **1973**, *95*, 4996

solubilizing agents than the more polar ones[42,43]. Small molecules such as p-xylene are taken up to a molar ratio of solute:deoxycholic acid of 2:1[44].

6

The most interesting result, however, transpires from the X-ray analysis of deoxycholic acid salts (see Figure 7.17)[45-48]. The crystals display a helical arrangement of the steroids which have a hydrophilic centre and a hydrophobic periphery. Several other methods indicate that the nonpolar face of sodium deoxycholate (= NaDC, **6**) in its micellar state in solution is also orientated towards the bulk aqueous medium. This was established by an NMR spectroscopic investigation of the sites of solubilization of aromatic probes using ring current-induced alterations of steroid proton chemical shifts. The effects were most profound on the 19-CH_3 groups[44,46], the outermost group of the assumed helix. The p-xylene/DCA ratio of 2:1 is more difficult to accommodate in the interior of a rigid micelle than on its surface. A ^{1}H- and ^{13}C-NMR study of the interaction between NaDC and a cholestane nitroxide probe also showed that the paramagnetic guest is bound to the hydrophobic face of NaDC. Again, space for the large spin labelled steroid should only be available on a convex outer surface but not on the concave inner side[47]. Small angle X-ray (SAXS)[48] and EXAF measurements[49] of NaDC solutions yielded the length and cross sectional radius of the micelle which was virtually identical to those found in crystals and fibres obtained by precipitation with acetone. NaDCA micelles also form strong complexes with the achiral bile pigment bilirubin in aqueous solution. Approximately 10% of bilirubin (wt/wt) could be dissolved and a CD spectrum ($\phi_{470} = -30000$)[50] was observed. This again points to a helical micelle structure (Figure 3.11) similar to the one observed in crystals.

None of the arguments on the micelle structure are conclusive, but all the evidence points to an ultrathin helical cylinder with at least two turns ($n = 12$) or at most 16 turns. This is also in accordance with the curvature of the NaDCA molecule and related A/B-*cis* steroids. Contrary to the molecules of classical detergents, bile salts have a lipophilic outer surface which is the convex side of the steroid nucleus and a hydrophilic inner surface, which is the polyhydroxylated concave side of the nucleus. It is a structurally satisfactory finding that the concave side of the cholic acid molecule also forms the concave inside of the micelle.

[45] F.M. Menger, J.-U. Rhee, L.J. Mandell, *J. Chem. Soc., Chem. Commun.*, **1973**, 918
[46] G. Conte, R. Di Blasi, E. Giglio, A. Parretta, N.V. Pavel, *J. Phys. Chem.*, **1984**, *88*, 5720
[47] G. Esposito, A. Zanobi, E. Giglio, N.V. Pavel, J.D. Campbell, *J. Phys. Chem.*, **1987**, *91*, 83
[48] G. Esposito, E. Giglio, N.V. Pavel, A. Zanobi, *J. Phys. Chem.*, **1987**, *91*, 356
[49] E. Giglio, N.V. Pavel, *J. Phys. Chem.*, **1988**, *92*, 2858
[50] R. Zana, D. Guveli, *J. Phys. Chem.*, **1985**, *89*, 1687

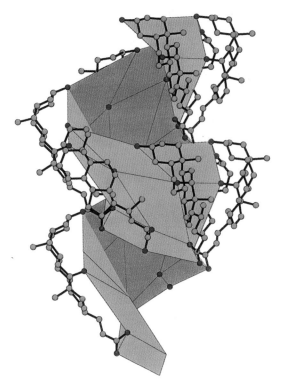

Figure 3.11 *Schematic model of a micelle consisting of 12 molecules of the rigid "facial" amphiphile deoxycholic acid. The polar surface is at the concave inside; apolar molecules are solubilized on the outer surface.*

3.9 Conversion of Micelles to Vesicles and Thin Films

Sodium and potassium salts of fatty acids (soaps) are the classical components of short lived micelles. Aggregation numbers are around 50–60; curvature is very high. Upon titration of clear alkaline micellar solutions of fatty acids with HCl, **translucence appears at a precise pH close to the pK_a,** varying according to temperature and chain length. Hydrogen bonds between COOH and COO⁻ groups were presumably formed, and planar bilayers appeared. If the temperature rose above a critical level, namely the melting point of the liquid-crystalline bilayer, vesicles were found in the dispersion. The most stable vesicles were obtained from oleate or 1:1 fatty acid–fatty alcohol mixtures[51–53]. Aggregation numbers were then in the order of 10^4; curvature was low.

The symmetrical bolaamphiphile **7** with two bipyridinium head groups, a long connecting diester chain and four bromide counterions dissolves in water.

---

[51] J.M. Gebicki, M. Hicks, *Nature (London)*, **1973**, *243*, 232
[52] J.M. Gebicki, M. Hicks, *Chem. Phys. Lipids*, **1976**, *16*, 142
[53] W.R. Hargreaves, D.W. Deamer, *Biochemistry*, **1978**, *17*, 3758

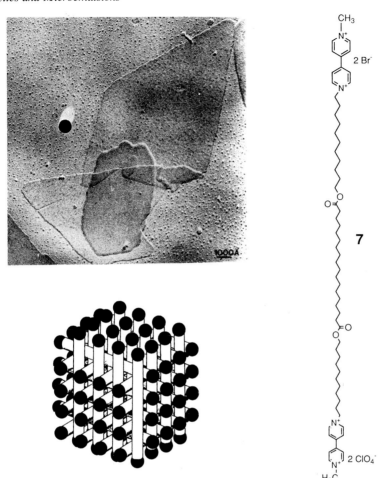

Figure 3.12 *Electron micrographs of* 7 nm *monolayers and micelles made of the bis-(bipyridinium)tetraperchlorate* 7. *A hypothetical model of the micelle is also given. The large sphere is a* 80 nm *latex reference.*

Electron micrographs show stable micelles of a diameter corresponding to the length of a molecule[54] (Figure 3.12). Models of these micelles invoked linear confrontations of the alkyl chains[54], but this has not as yet been verified. Upon titration of the micellar solutions with 1–2 moles of perchlorate, the solution became turbid and uniform vesicles were found. **One-sided precipitation of the bolaamphiphile** has been made responsible for this particular curved membrane structure formation (Figure 4.5). The same effect was obtained with a half-reduction of the bis-viologen bolaamphiphile (Figure 4.10). In both cases, only one half of the bolaamphiphile head groups merge together, whereas the other half remains soluble, or, rather, hydrated to a greater extent. Complete

[54] J.-H. Fuhrhop, D. Fritsch, B. Tesche, H. Schmiady, *J. Am. Chem. Soc.*, **1984**, *106*, 1998

replacement of bromide counterions by perchlorate or the quantitative reduction on both sides of the bolaamphiphile leads, as one would expect, to planar monolayers and massive amounts of stable bolaamphiphile micelles (Figure 3.12).

"$C_{18}N_3$" amphiphile **8a** forms 5 nm micelles in water with a cmc of 0.9×10^{-4} M. In contrast, "hyperextended" $C_{28}N_3$ **8b** aggregates at concentrations too low to be determined by surface tension or ^{81}Br NMR spectroscopy. Upon sonication, only vesicles were observed in this case[55].

8a

8b

Other conversion methods depend on the disruption of micelles containing water-insoluble compounds. For example, lipids e.g. lecithins can be reasonably well cosolubilized in micelles of sodium cholate or SDS. Dialysis slowly removes monomeric detergent molecules which dissociate away from the micelle resulting in a more concentrated lipid. Time and temperature controlled dialysis of the micelles finally yields monolamellar vesicles of uniform radii[56].

A somewhat different micelle-disruption method was used to prepare thin reactive films on electrode surfaces. For this purpose, micelles containing ferrocenyl surfactant **9** and a dissolved dye, e.g. phthalocyanine or quinone, were electrolyzed. The ferrocenyl micelles broke up into monomers whenever the iron atoms were oxidized electrochemically. The dissolved dyes then precipitated as transparent nanometre films onto the electrode surface[57].

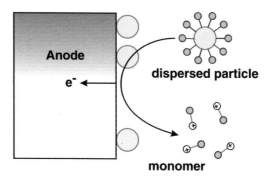

Figure 3.13 *Oxidation of the ferrocenyl surfactant **9** on the anode dissipates its micelles. Dispersed dyes precipitate as mono- and multilayers on the electrode's surface*[57].

[55] F.M. Menger, Y. Yamasaki, *J. Am. Chem. Soc.*, **1993**, *115*, 3840
[56] M.H.W. Milsmann, R.A. Schwendener, H.G. Weder, *Biochim. Biophys. Acta*, **1978**, *512*, 147
[57] T. Saji, K. Hoshino, Y. Ishii, M. Goto, *J. Am. Chem. Soc.*, **1991**, *113*, 450

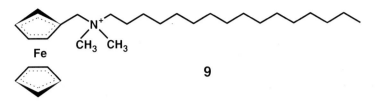

9

As a general result, we conclude that micelle and vesicle formation cannot be explained by "cone" or "cylinder" shapes of the monomeric amphiphiles. The key criterion for the curvature of molecular assemblies lies in the saturation solubility or cmc of the amphiphile. A cmc above 10^{-4} M usually means appreciable dissociation leading to small aggregation numbers of micelles. A cmc below 10^{-6} M means large planar bilayers or, upon their disruption, vesicles.

3.10 The Covalent Micelle-like Behaviour of Dendritic Macromolecules

Dendritic macromolecules are characterized by a large number of terminal groups all emanating from a central core. A large number of symmetrically arranged "branches" result in a three-dimensional globular shape. At high molecular weights, micellar-like spheres may result.

Several dendritic macromolecules were designed as covalent analogues of spherical micelles, but the majority are so tightly packed it is impossible for them to entrap dyes. Only the water-soluble unimolecular polyether micelle **10** was capable of dissolving 0.45 molecules of pyrene. A typical SDS micelle dissolves around 0.93 molecules per micelle. However, the SDS micelle with an aggregation number of 62 weighs twice the molecular weight of the dendritic polyether meaning that the solubilizing powers are very similar. The properties of the polymer micelles are, of course, independent of concentration.

No great extra effort was required in order to synthesize **polymers with one half-hydrophobic and one half-hydrophilic surface**, e.g. **11**. Such dendrimers are ideal for stabilizing water–dichloromethane emulsions over several weeks[58]. They are unique to covalent polymers; one cannot think of an equivalent molecular assembly in bulk solvents.

In industry, several dendrimers are already being synthesized on a hundred kilogram scale. The hopes for effective solubilizing systems suitable for medical applications may, however, not be justified. Phenolic ether or polyamine compounds are both normally not biodegradable so that major problems are foreseeable. An alternative is the employment of hydrolysable polyesters. One such compound already reported[59], is the imperfectly hyperbranched dendrimer which is easily accessible via thermal self-condensation of 3,5-bis(trimethylsiloxy)benzoyl chloride. This one-step procedure leads to a dendritic polyester

---

[58] C.J. Hawker, K.L. Wooley, J.M.J. Fréchet, *J. Chem. Soc., Perkin Trans. 1*, **1993**, 1287
[59] C.J. Hawker, R. Lee, J.M.J. Fréchet, *J. Am. Chem. Soc.*, **1991**, *113*, 4583

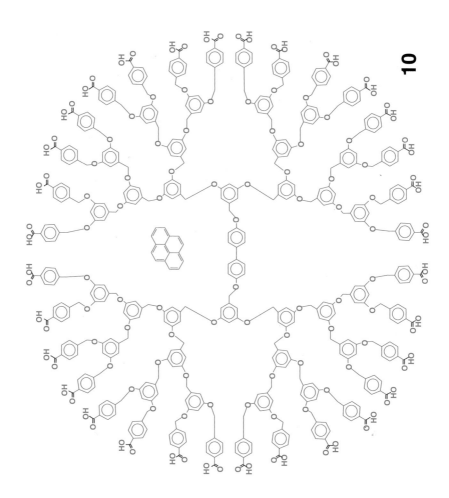

10

Micelles and Microemulsions

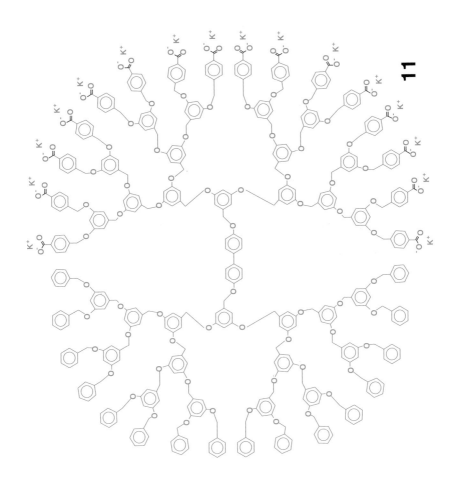

11

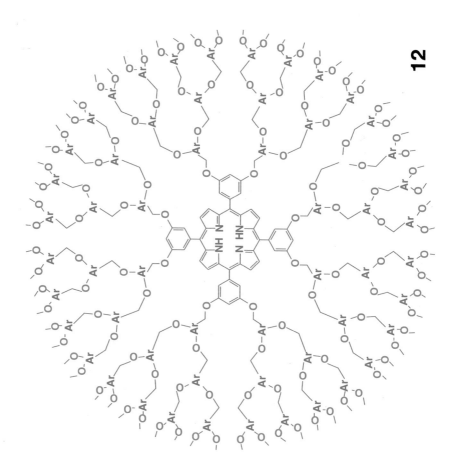

12

with molecular weights of about 10^5. The degree of branching was around 60%. No solubilization experiments have been reported so far.

The dendritic molecule **12** with an "integrated" porphyrin core and a catechol surface could be dissolved in acetonitrile. Its fluorescence was easily quenched by vitamin K_3, which presumably finds accommodation within the micellar core. Attempts to dissolve a porphyrin within the porphyrin micelle failed. They are too large and cannot interact with the dendritic porphyrin[60].

3.11 The Solubilization of Enzymes, Carbohydrates and Inorganic Colloids in Organic Solvents

We now turn to micellar systems which also contain organic solvents. Water pools present in organic solvents can be stabilized by amphiphiles and can be formed through the addition of water to heptane solutions of the sulfosuccinate AOT **14** (see Figure 3.14). The water pools, or inverted micelles, are structurally ill-defined, but one may safely assume that the sulfonate groups dip into the water regions whereas the alkyl chains lie in the aprotic solvent. A homogeneous mixture of 10% water in heptane can be prepared with only 0.1 M AOT. The water pool may be used to dissolve enzymes, e.g. chymotrypsin, which then catalyzes the hydrolysis of water-insoluble substrates, e.g. N-acetyl-L-tryptophan[61]. Enzyme activity is independent of pool size, which suggests that chymotrypsin molecules create their own micelles. N-glutaryl-L-phenylalanine p-nitroanilide is even more lipophilic and is rapidly degraded in isooctane–AOT–water mixtures[62] where it is possible to synthesize hydrocarbon-soluble peptides with enzymes in reverse micelles[63] (Figure 3.14).

The thio-disuccinate **13** aggregates in chloroform/cyclohexane (1:1) to form inverse micelles, which bind and solubilize carbohydrates. ^{1}H-NMR spectra of nitrophenolates and ESR spectra of TEMPO derivatives indicate that the carbohydrates bind tightly to the surfactant's head group at low water content[64].

Platinum salts were incorporated into water in an oil emulsion, e.g. pentaethylene glycol dodecyl ether in hexadecane/water (= inverted micelle) with well-defined cavities. The platinum colloids which are then produced by hydrazine hydrate are uniform. No measurable particles fell outside the limit of

---

[60] R.-H. Jin, T. Aida, S. Inoue, *J. Chem. Soc., Chem. Commun.*, **1993**, 1261
[61] F.M. Menger, K. Yamada, *J. Am. Chem. Soc.*, **1979**, *101*, 6731
[62] S. Barbaric, P.L. Luisi, *J. Am. Chem. Soc.*, **1981**, *103*, 4239
[63] P. Lüthi, P.L. Luisi, *J. Am. Chem. Soc.*, **1984**, *106*, 7285
[64] N. Greenspoon, E. Wachtel, *J. Am. Chem. Soc.*, **1991**, *113*, 7233

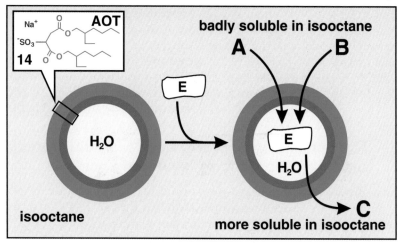

Figure 3.14 *Inverted AOT micelles dispersed in isooctane dissolve enzymes in the central water volume (inverse micelles). Polar substrates are enzymatically converted to less polar products. They are re-ejected into the apolar isooctane medium; the enzyme in the water pool can take up more substrate*[63].

25–35 Å in particle size distribution curves[65]. Colloidal gold, silver, platinum and platinized cadmium sulfide were generated in Aerosol-OT reversed micelles or in microemulsions by *in situ* photolysis of the appropriate ions (Figure 3.15)[66–68]. Under suitable conditions, each assembly contained approximately eight Au^{3+} ions, which firstly led to the formation of Au_8 clusters[67].

3.12 Fat Droplets and Microemulsions and their Application in Surface Cleaning and as Catalytic Systems

The act of cleaning the surface of a soiled material in a water bath containing dissolved detergents – simply called "washing" – signifies the most important application of amphiphilic molecules[69,70]. Most dirt is of an organic nature and originates from sebum, foods, synthetic oils and dyes, excrements and many

---

[65] M. Boutonnet, J. Kizling, P. Stenius, G. Maire, *Colloids Surfaces*, **1982**, *5*, 209
[66] K. Kurihara, J. Kizling, P. Stenius, J.H. Fendler, *J. Am. Chem. Soc.*, **1983**, *105*, 2574
[67] M. Meyer, C.L. Wallberg, K. Kurihara, J.H. Fendler, *J. Chem. Soc., Chem. Commun.*, **1984**, 90
[68] J.H. Fendler, *Chem. Rev.*, **1987**, *87*, 877
[69] Kirk–Othmer, Encyclopedia of Chemical Technology, 3rd ed., Wiley, New York, **1978**, see surfactants (Vol. 22), bleaching agents (Vol. 14) and enzymes (Vol. 19)
[70] Ullmann's Encyclopedia of Industrial Chemistry, 5th ed., VCH, Weinheim, **1986**, see bleaching (Vol. A4), cleansing agents (Vol. A7) and detergents (Vol. A8)

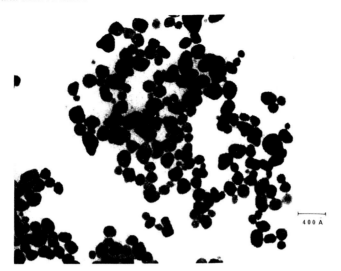

Figure 3.15 *Electron micrograph of colloidal platinum particles precipitated in inverse micelles*[66].

other biological or industrial sources. Inorganic dusts can usually be removed with a vacuum cleaner or a brush. Where the dirt is grease, anionic or neutral amphiphiles present in all practical washing systems will concentrate within the soft fat films; their surfaces become polar and can then be hydrated. The oily dirt "rolls-up" and forms water-soluble fat droplets (Figure 3.16). Ca^{2+} and Mg^{2+} counter-ions, which lead to irreversible precipitation, should be removed by "builders', such as EDTA, polyacrylate, sodium phosphate or with zeolites (most recommendable). Dirt redeposition in the rinsing stage is prevented by carboxymethyl cellulose additives, which presumably function by adsorption on to the dirt and the fabric, thus intensifying surface repulsions. Suspended fat droplets in washing water shrink (< 100 nm) because of their surface charge. Their emulsion is not a white milk, but resembles the slightly opaque suspensions of vesicles.

However, what applies to soft grease simply does not function with hard organic solids. Stacked multilayers of aromatic dyes on surfaces (e.g. tetra-sulfonated aniline blue of ink) are not separated by the alkyl chains of detergents. Instead they must be removed by oxidative bleaching, although the solid compounds are water-soluble as long as they are not adsorbed onto surfaces. Polar compounds such as haemoglobin also cannot be removed by detergents alone. Only enzymatic degradation or the action of molecules with very high dipole moments (urea derivatives) can assist here (Figure 3.16).

Wash water, of course, does not come into the category of supramolecular assembly chemistry, but several fundamental facts do appear here in the most simple form:

(i) Surfaces bind all kinds of organic molecules in thin layers.

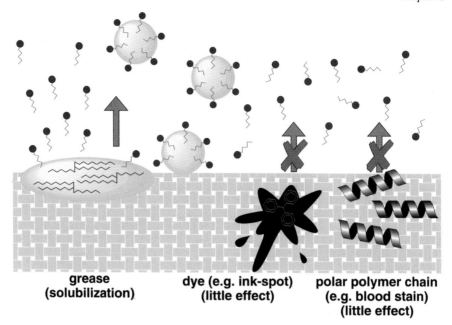

Figure 3.16 *The washing process of a dirty cloth. Only grease takes up amphiphilic molecules and is solubilized in water. Rigid dye assemblies and protein fibres are not removed, although their components were water-soluble before they were adsorbed on the cloth's surface.*

(ii) Water alone can neither remove water-insoluble nor water-soluble materials quantitatively from surfaces.

(iii) Solubilization with detergents is most effective for soft materials made of compounds with long alkyl chains (fats, fatty acids, hydrocarbons).

(iv) Planar dye molecules can often only be dissolved by destruction ("bleaching"), or by the use of detergent solutions above their cmc or by solvents.

Another type of structurally ill-defined emulsion between water and organic compounds can be applied as an efficient reaction medium for reactions between hydrophilic inorganic salts and hydrophobic organic substrates. So-called "microemulsions" are particularly useful in the case where the isolation of the reaction products is not necessary.

Microemulsions are isotropic and optically clear dispersions of hydrocarbons-in-water or water-in-hydrocarbons, where oil or water droplets are small (5–50 nm). Microemulsions are also thermodynamically stable and they remain clear indefinitely. They form spontaneously when water, hydrocarbon, surfactant and cosurfactant are mixed in specific proportions. Since microemulsions contain *no* defined supramolecular structures whatsover, they are of limited interest to organic chemists.

Nevertheless, oil-in-water microemulsions sometimes provide the medium for spectacularly efficient reactions which do not function in other environments. One important example is the destruction of the oil-soluble chemical warfare agent mustard. 5% aqueous hypochlorite (as found in domestic bleach) was added to half mustard, $EtSCH_2CH_2Cl$, which is much less dangerous than mustard but manifests similar chemical reactions, and then dissolved in 15 mL of microemulsion (82.1% H_2O; 3.2% cyclohexane; 4.9% SDS; 9.8% 1-butanol)[71]. The sulfide was quantitatively oxidized exclusively to the sulfoxide within 15 s. The exceptional speed of this reaction was explained through the formation of an alkyl hypochlorite at the oil/water interface (Figure 3.17).

Each of the six components in the "community of molecules" (water, hydrocarbon, surfactant, cosurfactant, oxidant and substrate) functions only by virtue of cooperative action, i.e. water acts as a solvent for the inorganic reagent; the cyclohexane droplets dissolve the substrate; both immiscible components must be combined with the mediation of the surfactant SDS; and the cosurfactant butanol fills the space between the charged SDS molecules. The result is, the droplets cannot grow and the emulsion becomes stable. There is a possibility that such microemulsions could work with several hydrophobic, environmentally contaminating materials and that the structurally

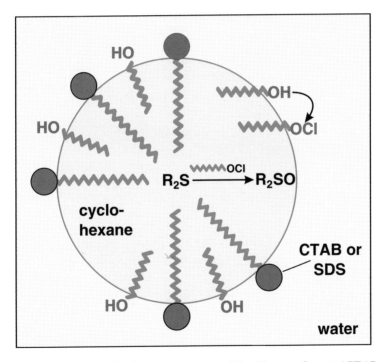

Figure 3.17 *Cyclohexane droplets in water are stabilized by a surfactant (CTAB; SDS) and a cosurfactant (dodecanol). The resulting microemulsion is very efficient in the oxidative destruction of half mustard ($R_2S = EtSCH_2CH_2Cl$) by sodium hypochlorite[71].*

[71] F.M. Menger, A.R. Elrington, *J. Am. Chem. Soc.*, **1991**, *113*, 9621

least-defined, self-assembling mixtures may turn out to be catalytically the most efficient. Several synkinons have been combined to form a perfect degradation system.

$$C_{12}H_{25}{}^{+}\overset{\overset{\displaystyle CH_3}{|}}{\underset{\underset{\displaystyle CH_3}{|}}{N}}\text{-}CH_2\overset{\overset{\displaystyle O}{\|}}{\underset{\underset{\displaystyle H}{|}}{C}}NPhNO_2 \qquad \mathbf{15}$$

Another interesting application of a structurally ill-defined assembly is the first efficient hydrolysis of an amide at room temperature and pH 8. When the cationic amide **15** (2×10^{-5} M) is mixed with anionic palmitate (2×10^{-5} M) at pH 8, an undefined molecular cluster is formed in which the water-stable amide ($t_{1/2} > 1$ yr) rapidly hydrolyzes ($t_{1/2} = 3.1$ min)[72]. The "long-time" contact of reactants, a state thought to be essential for enzyme-like reactions, can obviously not only be enforced by neighbouring group effects in rigid molecules or in complexes between enzyme clefts and substrates[73]. It may also occur in very simple micellar-like clusters of extremely low cmc.

---

[72] F.M. Menger, Z.X. Fei, *Angew. Chem.*, **1994**, *106*, 329; *Angew. Chem., Int. Ed. Engl.*, **1994**, *33*, 346
[73] F.M. Menger, *Angew. Chem.*, **1991**, *103*, 1104; *Angew. Chem., Int. Ed. Engl.*, **1991**, *30*, 1086

CHAPTER 4

Molecular Monolayer and Bilayer Vesicle Membranes

4.1 Introduction

Vesicles (Latin *vesicula*, bladder) are sealed, extremely thin (< 10 nm), often spherical membranes[1-4] which enclose aqueous or other solvent volumes of approximately 10^5–10^8 nm^3. Aggregation numbers are in the order of 10^4–10^5. The monomers are held together by the same solvophobic effects which produce micelles, and solvation forces together with membrane undulations prevent crystallization. The state of the vesicle membrane is therefore also essentially of a fluid character. Supramolecular ordering within vesicle membranes is negligible.

Nevertheless, synthetic vesicles can be organized because they are long-lived and have very low critical concentrations ($< 10^{-5}$ M). Few monomers are in equilibrium with the large assemblies. The lipid mono- and bilayers can therefore be used as follows:

(i) to entrap water-soluble dyes (e.g. calcein, sulfonated indigo), inorganic colloids (e.g. Pt°, MnOOH) and organic polyelectrolyes (proteins, DNA)
(ii) to locate proton or electron donating head groups on the outer surface and acceptors on the inner surface of a vesicle
(iii) to combine photoactive redox pairs of an organic and inorganic nature within the narrow space of a 2–6 nm thick membrane
(iv) to carry out the same recognition processes between chiral solutes and chiral membrane surfaces in bulk solution, which were established in 2D surface monolayers (see chapter 6).

Again, we leave the more physical aspects of vesicular bilayers to other text books[1-4] and concentrate on supramolecular structure and reactivity.

---

[1] J.H. Fendler, Membrane Mimetic Chemistry, Wiley, New York, **1982**
[2] G. Cevc, D. Marsh, Phospholipid Bilayers, Wiley, New York, **1987**
[3] J.N. Israelachvili, Intermolecular and Surface Forces, Academic Press, London, **1985**
[4] D.D. Lasic, Liposomes, Elsevier, Amsterdam, **1993**

4.2 The Self-Assembly of Vesicles and why they Remain in Solution

Firstly we have to differentiate between monolayer (MLM) and bilayer (BLM) lipid membranes in vesicles. **MLMs are composed of bolaamphiphiles**; these are amphiphiles which carry two head groups, namely one on each end of a hydrophobic core. Two head groups instead of one renders the amphiphile more water-soluble. **Two short alkyl chains** with 12 or more methylene groups, **or one long chain** with more than 24 hydrophobic atoms **must be employed** in order to obtain amphiphiles with a low critical vesicular concentration ("cvc"; $< 10^{-5}$ M). The general abbreviation "cmc" is, however, usually applied instead of "cvc".

The shortest and most easily accessible bolaamphiphiles which produce vesicles[4-9] with an aggregation number of about 10^4 upon sonication were produced from a disuccinic acid macrolide containing two 1,12-dodecanediol units[10,11]. The symmetric compounds **1a–c**, for example, form uniform vesicles upon sonication with an average diameter of around 30 nm and a membrane thickness of 2.0 nm, including the head groups. This means that the outer surface area of the vesicles is approximately 1.3 times larger than the inner surface area, although both head groups are identical.

The formation of a highly curved vesicle structure from a symmetric bolaamphiphile can only be understood if one assumes that in general, the amphiphiles tend to form extremely small spherical particles under thermally enforced equilibrium conditions. As a consequence, vesicles are formed from entropy effects (smallest possible assembly), solvophobic forces (assembly is enforced, because solvation forces are too weak for a monomolecular dissolution) and the solvation of head groups (crystallization is hindered). **No geometric effects are required in order to enforce the curvature of the vesicles**. The voids between the outer head groups most probably do *not* appear as empty spaces between the domain of parallel blocks (Figure 4.1b), as it was discovered that MLMs are profoundly efficient barriers for ions and electrons[12]. Small regular spacings (Figure 4.1a) which are filled presumably by a tilting of the hydrocarbon chains at the outer ends are most probable.

In extended bilayer lipid membranes ("Myelin figures"), made of single

---

[5] J.-H. Fuhrhop, J. Mathieu, *Angew. Chem.*, **1984**, *96*, 124; *Angew. Chem., Int. Ed. Engl.*, **1984**, *23*, 100
[6] J.-H. Fuhrhop, D. Fritsch, *Acc. Chem. Res.*, **1986**, *19*, 130
[7] J.-H. Fuhrhop, D. Fritsch, *Systematic Appl. Microbiol.*, **1986**, *7*, 272
[8] J.-H. Fuhrhop, S. Svenson, in Kinetics and Catalysis in Microheterogeneous Systems, (Surfactant Science Series, Vol. 38), M. Dekker Inc., New York, **1991**, 273
[9] J.-H. Fuhrhop, T. Bach, Advances in Supramolecular Chemistry 2, JAI Press, Greenwich, Connecticut, **1992**, 25
[10] J.-H. Fuhrhop, K. Ellermann, H.-H. David, J. Mathieu, *Angew. Chem.*, **1982**, *94*, 444; *Angew. Chem., Int. Ed. Engl.*, **1982**, *21*, 440
[11] J.-H. Fuhrhop, H.-H. David, J. Mathieu, U. Liman, H.-J. Winter, E. Boekema, *J. Am. Chem. Soc.*, **1986**, *108*, 1785
[12] E. Baumgartner, J.-H. Fuhrhop, *Angew. Chem.*, **1980**, *92*, 564; *Angew. Chem., Int. Ed. Engl.*, **1980**, *19*, 550

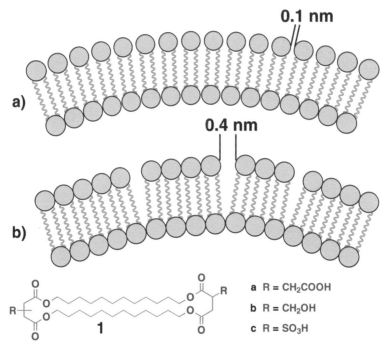

Figure 4.1 *Models of a 2D layer within a monolayer vesicle membrane made of bolaamphiphiles **1a–c** with identical head groups a) with regular spacings between head groups, b) with small domains. The distances given between the head groups relate to a vesicle circumference with an inner diameter of 25 nm, an outer diameter of 30 nm and a membrane thickness of 2.5 nm.*

headed amphiphiles, it is assumed that the sonication of planar bilayers removes amphiphile blocks (Figure 4.2a). The gaps are closed by surface diffusion of the water-insoluble amphiphiles and vesicle membranes with curvature result. The bilayer thickness usually approaches 4 nm. In a 30 nm wide vesicle, the outer surface is therefore larger than the inner surface by a factor of 1.8. This is the same ratio for the number of outer phosphate groups to the number of inner phosphate groups found in phospholipid vesicles with an approximate diameter of 30 nm as determined by ^{32}P-NMR spectroscopy[13]. This concedes that small bilayer vesicles contain almost double the amount of molecules in the outer layer than in the inner layer. This also means that almost twice as many outer oligomethylene chains meet the inside chains and that there must be some mobility around the methyl groups in the centre of the bilayer membrane.

DPPC vesicles (see Scheme 2.2) with an outer diameter of 22 nm show a total bilayer thickness of 3.5 nm, which is virtually identical to that of planar bilayers.

---

[13] a) J.A. Berden, R.W. Barker, G.K. Radda, *Biochim. Biophys. Acta*, **1975**, *375*, 186
  b) A. Chrzeszczyk, A. Wishnia, C.S. Springer jr., *Biochim. Biophys. Acta*, **1977**, *470*, 161

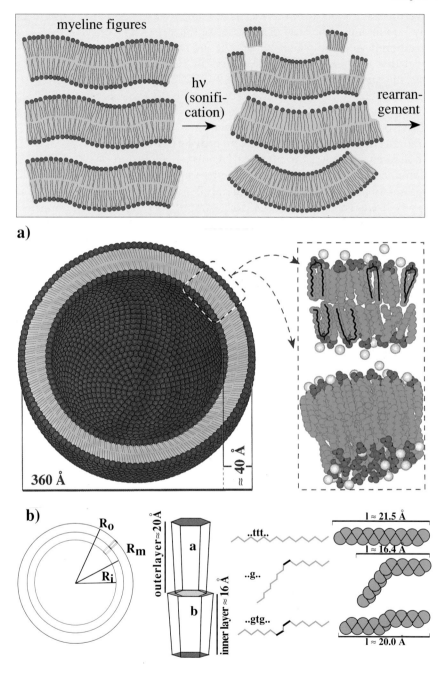

Figure 4.2 *a) The formation of a didodecyldimethylammonium bromide (DODAB) vesicle by sonication of planar bilayers (BLM).*
b) Schematic description of a vesicular DODAB membrane. In the liquid state the all-trans *conformation is mixed with several* $-60°$ *and* $+60°$ *gauche bends which produce kinks.*

Nevertheless, the outer monolayer is 2.0 nm, whereas the inner monolayer is only 1.5 nm thick. The head groups area of the inner surface is 0.68 nm^2, which corresponds with the planar value, but the inside tails are folded to such an extent that they decrease the radial lengths. The reverse applies to the outer layer: the surface per head group being 0.76 nm^2, tapering to 0.51 nm^2 in the tail region[14] (Figure 4.2b). Solutes can thus be localized either near the outer surface or near the centre of vesicle membranes.

Electron microscopy provides perfect pictures of vacuum collapsed vesicle membranes after negative staining with heavy metal salts. BLMs appear usually as collapsed balls, MLMs often as flat disks (see Figure 4.29). There is no requirement for double-chain amphiphiles in order to form vesicles. The same single-chain amphiphiles which form micelles also form vesicles if their charge is neutralized. This was practised, for example, via the protonation of soaps[14] or through addition of an amphiphilic counterion[15]. In both cases, fatty acids function perfectly well in the form of vesicles.

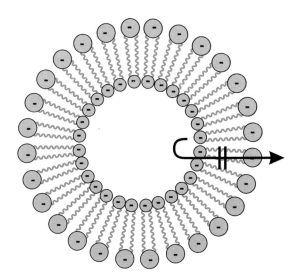

Figure 4.3 *Vesicular monolayer lipid membranes (MLM) made of bolaamphiphiles with a charge on the inner surface are stable, because charged head groups cannot cross the hydrophobic membranes. Dissociation and fusion of MLMs are slow.*

In the MLM and BLM vesicles, the cmc is small ($< 10^{-5}$ M). In MLMs, the solubility of the individual monomers may be relatively large, if they contain charged head groups. Once the bolaamphiphiles are entrapped in a vesicular assembly they cannot escape, as the polar head group would have to pass through an apolar membrane, which is an unlikely process. At pH = 9, for example, the diacetic acid **1a** dissolves reasonably well as the dianion in water.

---

[14] W.R. Hargreaves, D.W. Deamer, *Biochemistry*, **1978**, *17*, 3759
[15] a) J.-H. Fuhrhop, D. Fritsch, B. Tesche, H. Schmiady, *J. Am. Chem. Soc.*, **1984**, *106*, 1998
   b) J.-H. Fuhrhop, D. Fritsch, *J. Am. Chem. Soc.*, **1984**, *106*, 4287

Upon acidification, vesicles are formed which do not dissociate into monomers[11] at pH = 9 (Figure 4.3). In BLMs such behaviour was not observed. Their monomers must always be of a low solubility to ensure that the vesicles do not rearrange to micelles, planar bilayers (Figure 4.2) or 3D crystals.

The typical aggregation number of the smallest possible bilayer vesicle (diameter: 22 nm) made of double-chain, egg-yolk lecithin[17] is approximately 2.5×10^3; and 1.0×10^4 for a 30 nm MLM vesicle with bipyridinium head groups. The molecular mass of such small, monolamellar vesicles is then around $1-3 \times 10^6$ Daltons. From the huge molecular mass alone, one would expect extremely strong van der Waals forces between encountering vesicles. Rapid precipitation of vesicles, however, does not usually occur, a fact which has been traced back to thermal fluctuations ("undulations", Figure 4.4) of the ultrathin fluid membranes[18,19]. Should vesicles meet, they continually "undulate" and separate. Only in the case where membranes are particularly stiff (e.g. by cooling down or by adding rigid solutes like cholesterol) do they precipitate rapidly.

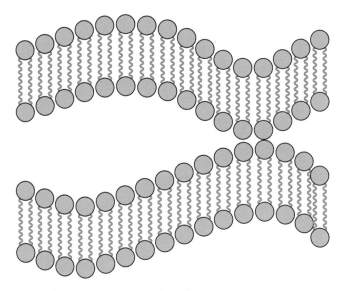

Figure 4.4 *Thermal undulations of vesicular MLMs (and BLMs) weaken intervesicular van der Waals interactions and prevent their aggregation and precipitation.*

Micellar solutions are long-lived because repulsion between the hydrated head groups overcomes the (weak) crystallization tendency of the solvophobic part of the molecule. The micelles themselves do not aggregate as they are too short-lived. Vesicles survive, as far as the monomer crystallization is concerned,

[16] A. Chrzeszczyk, A. Wishnia, C.S. Springer jr., *Biochim. Biophys. Acta*, **1977**, *470*, 161
[17] C. Huang, J.T. Mason, *Proc. Natl. Acad. Sci. USA*, **1978**, *75*, 308
[18] W. Helfrich, *Z. Naturforsch.*, **1973**, *28c*, 693
[19] M.D. Mitor, J.F. Faucon, P. Météard, P. Bothorel, *Adv. Supramol. Chem.*, **1992**, *2*, 93

for the same reasons. The awaited precipitation of vesicle clusters, however, is not prevented by rapid disassembly, but via mechanical undulations.

4.3 Synkinesis of Asymmetric Vesicle Membranes (In–Out)

A topological detail unique to vesicle membranes lies in the possibility of arranging several thousands of head groups A on the outer surface of the membrane and a similar number of head groups B on the inner surface. This is particularly interesting if B is an electron donor and A an acceptor, or vice versa. A membrane-dissolved dye may then, in its excited state, donate electrons to A and be re-reduced by B. $A^{\cdot -}$ and $B^{\cdot +}$ radicals on both sides of the membranes would then be formed and could possibly induce further chemical redox reactions, e.g. the splitting of water (see sections 3.7 and 4.4).

In nature, asymmetry is achieved through membrane dissolved proteins. In lipid membrane systems without proteins, only monolayers made of bolaamphiphiles allow a totally asymmetric arrangement of head groups. The simplest asymmetry to be achieved is dependent on the one-sided precipitation of bolaamphiphiles. α,ω-Dicarboxylic acids, for example, are often soluble at pH > 8 and spontaneously form vesicles upon acidification to pH 5. At a lower pH, all carboxyl groups become protonated and one usually observes ill-defined precipitates[11].

The same behaviour can be observed for α,ω-bis(dipyridinium) salts, e.g. **2**. The tetrabromide is easily soluble in water, but upon addition of 1–2 moles perchlorate or iodide, a "one-sided precipitation" occurs[15], meaning that the insoluble bipyridinium perchlorate head groups move together, whereas the bipyridinium bromide head groups stay more hydrated. The curvature of extremely uniform vesicle membranes is thus induced (Figure 4.5) and the perchlorate or iodide ions are entrapped within the vesicle. Vesicle formation

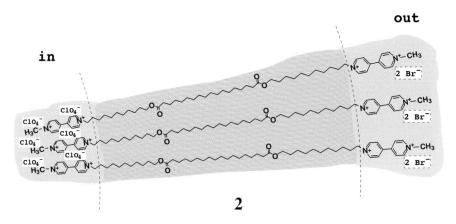

Figure 4.5 *A small part of a monolayered vesicle membrane made by a one-sided precipitation of the given bipyridinium bolaamphiphile 2 with perchlorate.*

from the water-dissolved tetrabromide also occurs when partially photoreduced in the presence of ferrocyanide. The viologen cation radical is once more less water-soluble than the dication, resulting in the gathering of the photoreduced sites on the inner surface[15a]. The same vesicles with viologen head groups bind benzidine molecules in the outer head group region and form polymeric, blue charge transfer complexes at temperatures below 20°C. Above 35° the complex dissociates and the vesicle decolorizes. This reaction is fully reversible and not accompanied by radical formation[15b].

Unsymmetrical monolayer lipid membranes (MLMs) were obtained in two cases **where bolaamphiphiles had one large head group and one small head group**. More than 99% of the small sulfonate head groups in **3** are located on the small inner surface, with the large succinate head group on the large outer surface of vesicle membranes. Proof of this asymmetric arrangement came from quantitative comparisons of the "metachromatic effect" of surface-adsorbed acridine orange[20]. The cationic dye aggregates in presence of anionic polyelectrolytes and a short-wavelength shift occurs. In mixed bilayer vesicles containing 1% sulfonate amphiphiles, the short-wavelength shift of acridine orange added to bulk water was easily detectable at pH 4, whereas the asymmetric bolaamphiphile **3** vesicles induced no shift because only electroneutral thiosuccinic acid head groups were present at the outer surface (Figure 4.6)[11,20]. In the case of the quinone bolaamphiphile **4** vesicle, it was shown that the quinone could be quantitatively reduced by external sodium borohydride, which does not penetrate vesicle membranes[21].

The **immiscibility of CF_2- and CH_2-chains** was also utilized for the preparation of unsymmetric vesicle membranes. The hydrophobic parts of bolaamphiphile **5** with fatty acid and fluorocarbon sulfonate halves do not mix and all the fluorosulfonate halves were on the outer side of the monolayered vesicle membrane (Figure 4.7)[22]. The sulfonate head group was once more localized by the metachromatic effect, the hydrophobic parts with F- and H-substituted spin labels.

With bilayer lipid membranes it is not possible to achieve a fully asymmetric arrangement of head groups or chains. There is no apparent reason why all the molecules of two independent layers should only concentrate in one layer. Nevertheless, a little asymmetric distribution is found in vesicles made of lipid mixtures. Cerebroside sulfate, an anionic monoglycosyl ceramide was, for example, added exclusively to the outer surface of a performed DPPC vesicle[23] (see Scheme 2.2) which was quantitized by the metachromatic effect of acridine orange.

A mixture of negatively charged phosphatidic acid and phosphatidylcholine was cosonicated. This system was used to quantify the metachromatic effects of

[20] J.-H. Fuhrhop, J. Mathieu, *J. Chem. Soc., Chem. Commun.*, **1983**, 144
[21] J.-H. Fuhrhop, H. Hungerbühler, U. Siggel, *Langmuir*, **1990**, *6*, 1295
[22] K. Liang, Y. Hui, *J. Am. Chem. Soc.*, **1992**, *114*, 6588
[23] B. Cestaro, E. Pistolesi, N. Hershhowitz, S. Gatt, *Biochim. Biophys. Acta*, **1982**, *685*, 13

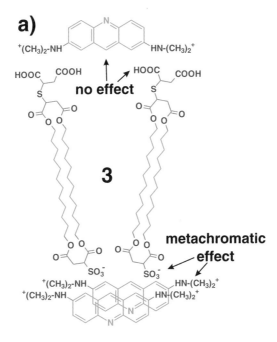

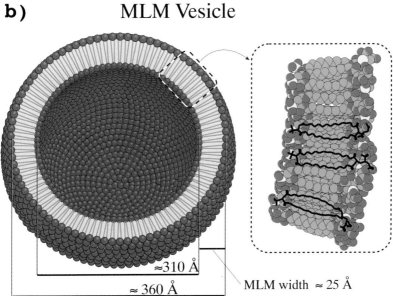

Figure 4.6 *a) The bolaamphiphile **3** with a large and a small head group assembles to form asymmetric vesicle membranes.*
b) The desired uniform distribution of charged and uncharged head groups on the outer and inner vesicle surface can be proven with added acridine orange. A short-wavelength shift is observed, if the membrane surfaces contain negative charges. As little as 1% of negatively charged head groups in an electroneutral membrane surface can be detected unequivocally.

acridine orange and methylene blue[24]. It was found that (i) phosphatidic acid molecules tend to be associated, in spite of the electrostatic repulsion between electronegative groups, (ii) the charged component is preferentially distributed into the external layer at a high ratio of phosphatidylcholine to negatively charged phospholipid and (iii) it is preferably located with the inner segment, if the electronegative lipid is in excess[25]. The latter, very surprising finding is thought to relate to the smallness of the phosphatidic head groups as compared to the choline head group. In the case of such double-chain amphiphiles which cannot leave the vesicle membrane because they are water-insoluble, steric effects thus also become important. The bolaamphiphile phenomenon discussed above also appears in bilayers; the deviations of experimentally determined distributions from statistical values, however, never exceed 30%.

Reaverage time, or so-called "flip-flop time", is important in respect to membrane asymmetry. It measures the time taken for an amphiphile to exchange

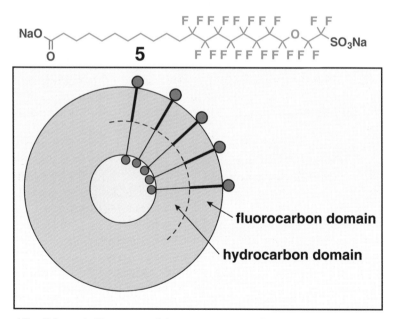

Figure 4.7 *Schematic illustration of the asymmetric vesicle MLM made of 5 with uniform hydrocarbon and fluorocarbon domains.*

[24] S. Massari, D. Pascolini, *Biochemistry*, **1977**, *16*, 1189
[25] S. Massari, D. Pascolini, G. Gradenigo, *Biochemistry*, **1978**, *17*, 4465

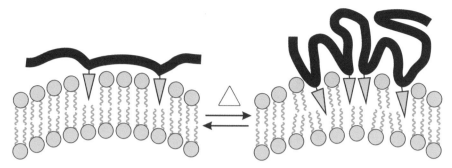

Figure 4.8 *Comb-shaped polyacrylate with hydrophobic side-chains integrates vesicle membranes, but hardly disturbs its ordering. Only melting of the membrane and/or neutralization of the acrylate carboxylate group produces strong local disturbances. The polymer now promotes lipid flip-flops[26].*

between the inner and outer monolayer. The lower limit of flip-flop rates t_{ff} is set by fatty acids, where the half-reequilibration rate t_{ff} is ~ 5 s. In the case of phospholipids, it rises to t_{ff} ~ 8 h and sinks to zero for membrane proteins[26]. Polyacrylates with a few hydrophobic side chains disturb the membrane surface in some phospholipid vesicle membranes and they act as "flippases"[27] (Figure 4.8). t_{ff} decreases from 12 h to < 1 h. Eventual equilibrium asymmetry of bilayer membranes remains, however, constant.

Covesicles of the cationic nitrobenzoate **6** and DODAC, or corresponding DPP-analogues, e.g. **7**, are hydrolysed at pH 8. Nitrophenolate absorption appears at 400 nm. The outer benzoate esters at the outer vesicle surface are hydrolysed within minutes and the same head groups on the inner surface survive for 1–15 hours (Figure 4.9). Detailed kinetics of flip-flop dynamics and OH^- permeation have been evaluated in these systems[28]. Monolayer lipid membranes made of macrocyclic bolaamphiphiles showed enhanced dynamic stability[29].

For periods shorter than about 6 h a symmetric bilayer lipid membrane can turn totally **asymmetric in respect to head group distribution by surface reactions** of vesicles. We give three examples (Figure 4.10). Phenylenediamine derivatives with two long alkyl chains on only one nitrogen atom, for example *N,N*-dioctadecylphenylenediamine, form stable vesicles upon sonication. A reaction with water-soluble ionic heterocycles only occurs on the outer surface and is usually of a quinonoid character, whereas the inside of the membrane retains its reductive phenylenediamine surface[30]. One may also replace the primary amine group of phenylenediamine by an azonium group, whereby the vesicles' surface is polycationic. Upon irradiation with visible light, the diazonium chloride

[26] E. Sackmann, *Ber. Bunsenges. Phys. Chem.*, **1978**, *82*, 891
[27] S. Bhattacharya, R.A. Moss, H. Ringsdorf, J. Simon, *J. Am. Chem. Soc.*, **1993**, *115*, 3812
[28] a) R.A. Moss, S. Bhattacharya, S. Chatterjee, *J. Am. Chem. Soc.*, **1989**, *111*, 3680
    b) R.A. Moss, Y. Okumura, *J. Am. Chem. Soc.*, **1992**, *114*, 1750
[29] R.A. Moss, G. Li, J.-M. Li, *J. Am. Chem. Soc.*, **1994**, *116*, 805
[30] J.-H. Fuhrhop, H. Bartsch, *Liebigs Ann. Chem.*, **1983**, 802

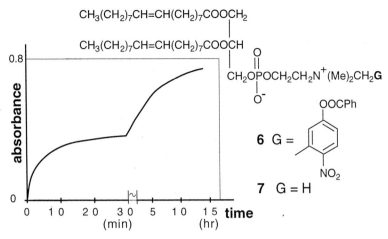

Figure 4.9 *Hydrolysis of phenol esters bound to the outer surface of phospholipid vesicles is finished after 3 minutes. The same reaction on the inner surface takes hours. Flippases (see Figure 4.8) accelerate the latter reaction.*

head group degrades and is replaced by chloride. The vesicle precipitates if too many benzene diazonium head groups are replaced by chlorobenzene. Since deazotization is a chain process one can, in principle, precipitate a vesicle containing approximately 10^4 molecules with a single light quantum. Photographic processes without the need of silver can be envisaged on this basis[30,31]. Redox-active bipyridinium head groups in MLM vesicle membranes are also only reduced on the outer surface if ionic reductants are applied in the bulk phase (Figure 4.10).

Nitrogen-containing head groups have also been developed which can reversibly take up protons or metal ions. The ylide **8a** forms yellow (λ_{max} = 399 nm) vesicles which become colourless, (**8b**, λ_{max} = 256 nm) on addition of acid. The bipyridinium derivative **9** is red (λ_{max} = 470 nm). As these compounds are water-insoluble, they are also suitable for incorporation into foreign vesicles in which the positively charged pyridinium residue lies next to the head groups of the host membrane. The pK_a for protonation of the ylide **8a** is very strongly influenced by the charge of the host groups[32]. In electroneutral vesicles, the pK_a value lies close to 6, in electropositive vesicles it is 4 and in electronegative vesicles it is 8.

The vesicle-forming ethylenediaminediacetic acid derivative **10** complexes metal ions (e.g. Cu^{2+}) from the aqueous phase. Since metal ions cannot generally penetrate hydrophobic vesicle membranes, **literally any redox-active ions can be fixed to the inner and/or outer membrane surfaces**[33]. Cationic metallovesicles were also obtained from the double chain amphiphile **11** synthe-

---

[31] J.-H. Fuhrhop, H. Bartsch, D. Fritsch, *Angew. Chem.*, **1981**, *93*, 797; *Angew. Chem., Int. Ed. Engl.*, **1981**, *20*, 804
[32] J.-H. Fuhrhop, G. Penzlin, H. Tank, *Chem. Phys. Lipids*, **1987**, *43*, 147
[33] J.-H. Fuhrhop, V. Koesling, G. Schönberger, *Liebigs Ann. Chem.*, **1984**, 1634

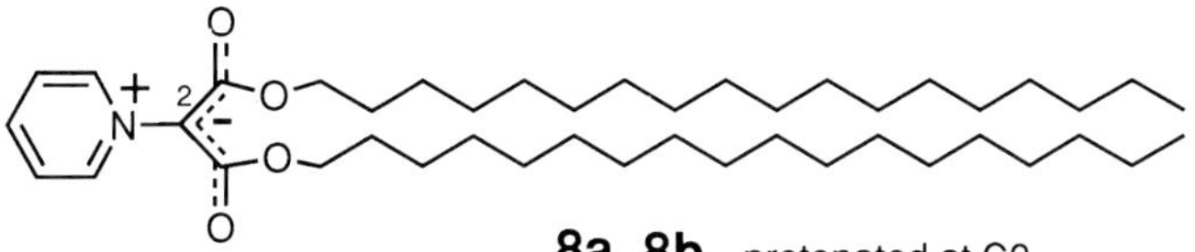

8a, 8b protonated at C2

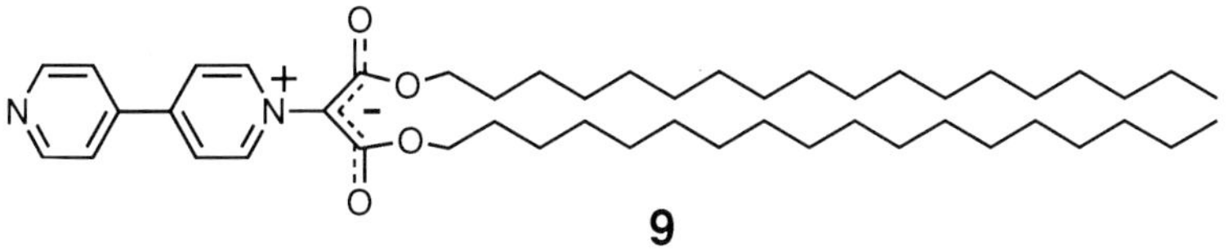

9

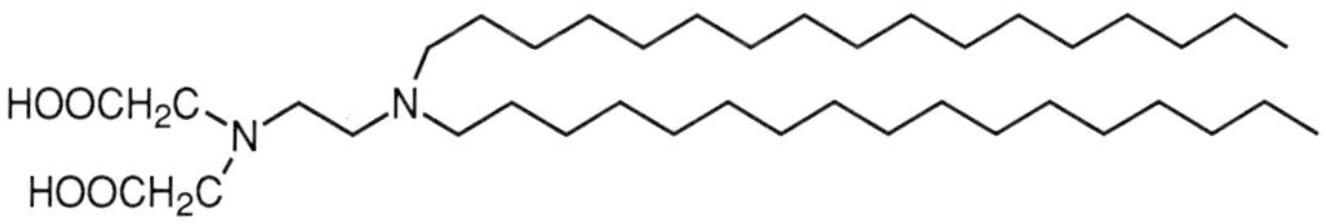

10

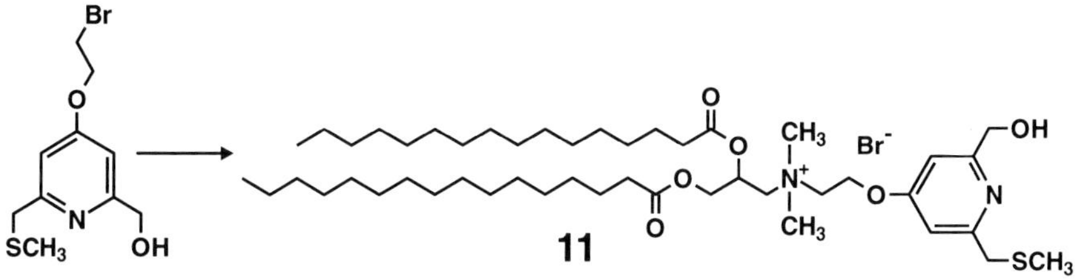

chelidamic acid

11

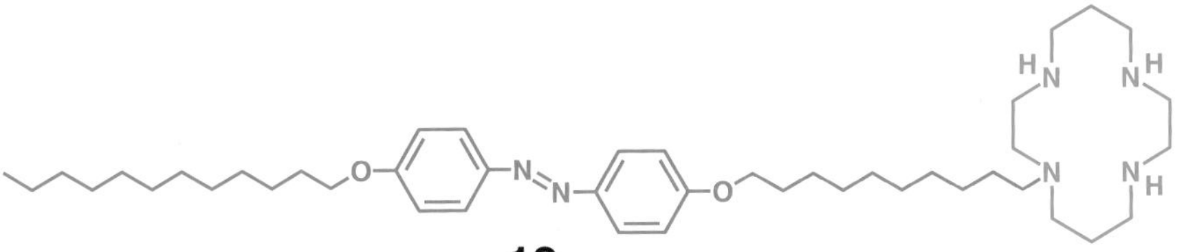

12

sized from commercial chelidamic acid. The vesicles were active as catalysts of the hydrolysis of nitrophenyl esters[34]. A single-chain amphiphile **12** with a cyclam head group and an azobenzene chromophore in the alkyl chain[35] was dissolved in dihexadecyldimethylammonium vesicles (1:1). At room temperature, the absorption maximum occurred at 326 nm and shifted to 356 nm above 27.3°C. Cu^{2+} bound to the vesicles and stabilized this H-aggregate up to 45°C. Unilamellar vesicles with diameters of 80–1200 Å were also obtained from tertiary diamides with crown-ether head groups[36], but there were no reports on

[34] P. Scrimin, P. Tecilla, U. Tonellato, *J. Am. Chem. Soc.*, **1992**, *114*, 5086
[35] A. Singh, L.-I. Tsao, M. Markowitz, B.P. Gaber, *Langmuir*, **1992**, *8*, 1570
[36] S. Munoz, J. Mallén, A. Nakano, Z. Chen, I. Gay, L. Echegoyen, G.W. Gokel, *J. Am. Chem. Soc.*, **1993**, *115*, 1705

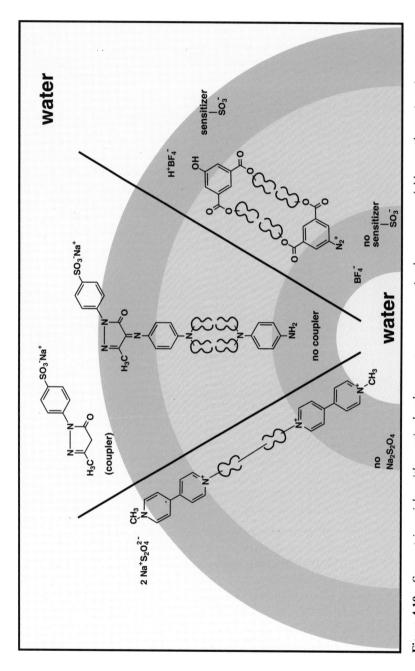

Figure 4.10 *Symmetric vesicles with reactive head groups turn asymmetric when a water-soluble, membrane-inactive reagent reacts only with the outer surface. Flip-flop usually takes hours and can be completely suppressed in MLMs made of charged bolaamphiphiles.*

metal ion effects. The conformation of the enormously large head groups is also unknown.

Carbohydrate head groups on the surface of vesicle membranes are recognized by lectins[37]. Concanavalin A (= Con A), for example, is a tetrameric protein with four binding sites for D-gluco- and D-manno-pyranosides[38]. The mixing of 10^{-4} M aqueous suspensions of vesicular glycolipids (Figure 4.12) with Con A (1 mg/mL), for example, causes immediate agglutination and precipitation of the vesicle–Con A adduct (Figure 4.11). Additions of 10^{-3} M D-methylglucoside solutions redissolve the precipitate. Vesicles with L-configured glucose or mannose, or D-configured galactose head groups give no precipitation[39]. The finding that the high stereoselectivity of action of Con A is also present with the corresponding tetraacetates was quite surprising[39]. **The protein obviously clearly distinguishes between two enantiomers, but fails to recognize the difference between four small, polar OH groups and four large, apolar OCOCH$_3$ groups**. Essentially planar binding sites on both the Con A and vesicle surfaces were proposed. From an estimated 23 kJ mol^{-1} binding energy, a protein–vesicle distance of about 0.45 nm was estimated using the

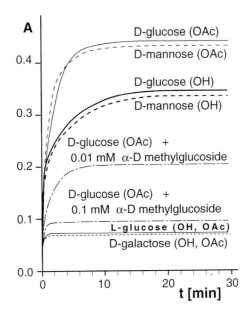

Figure 4.11 *Rise of absorption (A) at 400 nm after addition of Con A to bilayer vesicles with carbohydrate head groups. D-glucose and D-mannose head groups are precipitated as well as their tetraacetates. D-galactose and L-glucose vesicles remain in solution, both as free carbohydrates and tetraacetates. 0.01 mM α-D-methylglucoside impedes the stereoselective action of Con A.*

[37] I.J. Goldstein, C.E. Hayers, *Adv. Carbohydr. Chem. Biochem.*, **1978**, *35*, 128
[38] H. Bader, K. Dorn, B. Hupfer, H. Ringsdorf, *Adv. Polym. Sci.*, **1985**, *64*, 1
[39] J.-H. Fuhrhop, M. Arlt, *Angew. Chem.*, **1990**, *102*, 699; *Angew. Chem., Int. Ed. Engl.*, **1990**, *29*, 672

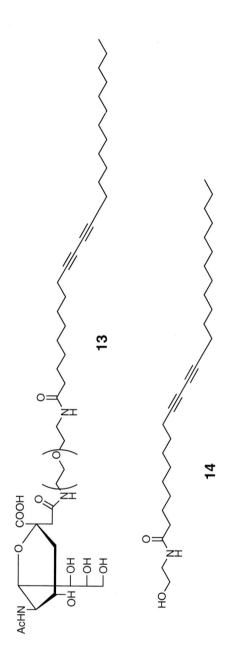

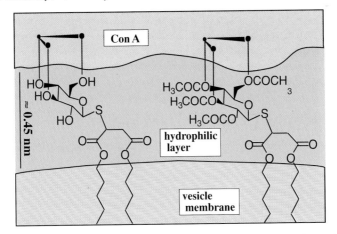

Figure 4.12 *The glucose-binding sites for 03, 04 and 06 are assumed to be on the β-pleated sheet surface of Con A with a vesicle–Con A distance of 0.45 nm. This would be consistent with the found binding constants as well as with the non-differentiation between OH groups (left side) and OCOCH$_3$ groups (right side).*

point dipole model for six dipole–dipole interactions. This distance leaves ample room for acetyl groups as well as hydration spheres (Figure 4.12).

The surface lectin of the influenza virus, haemagglutinin, binds to terminal α-glycosides on cell-surface glycoproteins and glycolipids. Polymerized liposomes (ellipsoids; 40 × 15 nm) containing 1% of the C-glycoside lipid **13** ("sialoside") and 99% of **14** were tested for binding to the influenza virus. To achieve 100% inhibition as little as 5.7×10^{-7} M concentration of the sialoside **13** was required, making it one of the most potent synthetic inhibitors. The vesicular preparation is approximately 30 000 times more efficient than the corresponding monovalent sialic acid derivatives[40].

Several pharmaceuticals were attached to the outer surface of vesicles by a labile bond. Vesicles containing 2,3-dipalmitoyl-sn-glycerol-1-phospho-3'-azidothymidine* **15**, for example, showed a greatly enhanced inhibition of the

---

* sn = "stereospecifically numbered" glycerol unit; headgroup at C-1 (see *Biochim. Biophys. Acta*, **1981**, *650*, 21)
[40] W. Spevak, J.O. Nagy, D.H. Charych, M.E. Schaefer, J.H. Gilbert, M.D. Bednarski, *J. Am. Chem. Soc.*, **1993**, *115*, 1146

human immunodeficiency virus (HIV) replication *in vitro*[41]. Such liposomes with inhibitors displayed on the surface could either interact with targeted surface receptors via a multivalent contact or hydrolyzed *in vivo* by cellular lipases and phospholipases[42]. The same phospholipase D was also applied in synthesis (see Scheme 2.3).

4.4 Membrane Dissolved Dyes and Steroids

Aqueous vesicle solutions display nine regions or compartments (Figure 4.13):

1. the outer water volume (bulk volume)
2. the aqueous volume in the vicinity of the head groups
3. the outer head groups
4. the hydrophobic membrane close to the outer head groups
5. the interior of the hydrophobic membrane (in bilayer membranes this region can be subdivided into an inner and outer half)
6–9. the inner regions corresponding to the outer regions 4–1; (9 is called the entrapped water volume)

Assuming an average microvesicle with an outer diameter of 50 nm, a membrane thickness of 3 nm and a concentration of 10^{-3} M lipid molecules corresponding to a 10^{-7} M vesicle solution, one obtains volumes of around 3 mL each of entrapped water and hydrophobic lipid membrane per litre of bulk water.

Due to their encyclopaedic characterization in the literature, we will begin with the physical behaviour of phospholipid bilayers[1-3]. Bilayers of DPPC exhibit two, first-order phase transitions, the so-called main transition at 42°C and the pretransition at 33°C. At the main transition temperature T_2, the *all-trans* configured lipid chain undergoes a transition to a fluid phase; at the pretransition temperature T_1, a rearrangement of the hydrated head groups occurs. In electron micrographs one sometimes observes a "ripple" structure of DPPC vesicles if they are cooled to a temperature below T_1. In chemical terms, one would also expect that at T_1 the specific binding properties of vesicle surfaces should change, but to the best of our knowledge, this has never been demonstrated. At T_2 one awaits a drastic increase of the solvation power of the fluidized membrane, a state which has been verified several times. This result is, however, of a relatively low degree of practical interest, since bilayer lipid membranes can be constantly kept in a fluid state by using amphiphile mixtures. For biological lipid membranes, the esterification of glycerol with fatty acids of differing chain lengths and degrees of unsaturation takes over this function. Biological vesicle membranes are therefore in a permanent liquid state at elevated or low temperatures. Vesicle membranes possess approximately the

---

[41] G.M.T. Wijik, F.W.J. Gadella, K.W.A. Wirtz, K.Y. Hestetler, M. von den Bosch, *Biochemistry*, **1992**, *31*, 5912
[42] P. Wang, M. Schuster, Y.-F. Wang, C.-H. Wong, *J. Am. Chem. Soc.*, **1993**, *115*, 10487

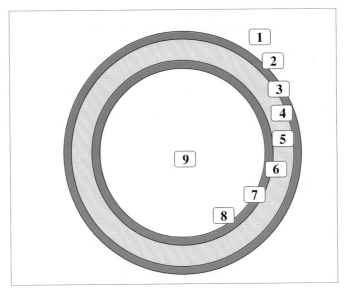

Figure 4.13 *Aqueous dispersions of vesicles embrace one large bulk volume of water (region 1) and 8 vesicular regions with different spatial relationships to region 1 and different polarities. See text.*

same solvation power as chloroform, meaning that a substance which can be dissolved in 3 mL of chloroform, can usually also be dissolved in 3 mL of a vesicle membrane. The only problem remaining is that these three mL are usually dispersed in 1 L of water.

The low concentration of vesicles ($\sim 10^{-3}$ M) and the even lower concentration of dissolved molecules make it necessary to label solutes with radioactive isotopes or to use dyes with strong absorption or fluorescence bands. [4-^{14}C]-β-sitosterol **16a** and cholesterol **16b**, are two steroids present in soy beans and egg yolk which differ only by an ethyl group. They were dissolved up to 50 mol % in DPPC vesicles at 50 °C. The steroids were then extracted from the vesicles with a lysophospholipid-enriched dispersion at 37 °C, whereby it was discovered that sitosterol was desorbed much more slowly than cholesterol by a factor of 4. The corresponding "dietary finding" by food chemists is that the intestinal absorption of sitosterol is much slower than that of cholesterol. It is therefore fully reflected in the phospholipid vesicle–lysophospholipid dispersion model system[43]. **The extra ethyl group obviously "fixes" sitosterol in membranes**. This could also relate to the extreme efficiency of norgestrel, the gestagen present in most contraceptive pills, which is much more potent than the natural analogue, carrying two methyl groups instead of one ethyl group.

The permeability of vesicle membranes to glucose is decreased by cholesterol. This steroid stiffens the membrane, making it more hydrophobic. β-Sitosterol has a minimal influence. **The extra ethyl side-chain also has the effect of**

[43] C.-C. Kan, R. Bittman, *J. Am. Chem. Soc.*, **1991**, *113*, 6650

loosening the membrane allowing water to enter. Fluorescence probes indicate a "more polar" vesicle membrane in the presence of β-sitosterol than in the presence of cholesterol, whereby glucose permeability increases correspondingly. Steroids with alkyl chains on C-17 reduce glucose leakage in DPPC vesicles, cholesterol being the most effective[44].

16 a R = Et (sitosterol)
16 b R = H (cholesterol)

Porphyrins are ideal for investigations of light-induced charge separation and oxygen transport in vesicular solutions. Their intense Soret band at 400 nm allows measurement at concentrations down to 10^{-6} M; their fluorescence spectra being detectable even at 10^{-8} M. Photo- and redox chemistries are extremely variable depending on the central metal ion and substituents[45a].

Water-soluble, charged porphyrins can be efficiently attached to vesicle surfaces. *Meso*-tetrapyridiniumporphyrin, for example, sticks to anionic vesicles (see Figure 4.16b)[45b]; β-octaacetic acid porphyrin **17** can be fixated to cationic vesicle surfaces. An interesting pH effect was observed[5] from the octaacetic acid porphyrins **17**. At pH 7 the porphyrin was dissolved in water as an oligoacetate anion. The longest wavelength absorbance band (α-band) occurred at 616 nm. In the presence of electroneutral or negatively charged vesicles, the porphyrin remained in the bulk water phase and upon gel chromatography, it separated from the vesicle fraction, something that does not happen in solutions of cationic vesicles. The visible spectrum of the adsorbed porphyrin also contains a 616 nm α band. Should the aqueous solution of **17** be acidified (pH 4.5), the porphyrin precipitates. In the presence of vesicles with surface charge, precipitation does not occur and the vesicles take up the porphyrin. If the pH is raised to 5.2, the visible spectrum of the neutral porphyrin chromophore changes to the three-banded spectrum of the porphyrin monocation, i.e. one of the pyrrolenine nitrogen atoms is protonated. This porphyrin cation is most probably dissolved in the aqueous regions 1 or 2 (see Figure 4.13). Approximately half of the acetate substituents should be protonated. At pH 5.0, the three-band porphyrin cation spectrum is replaced by the four-band spectrum of the non-protonated porphyrin base. The α band is now to be found at 625 nm, corresponding to the spectrum of porphyrin octaacetic acid in chloroform. It has been concluded from the spectral shift that the porphyrin "migrated" from the aqueous medium to the hydrophobic membrane region 4.

A protonated nitrogen base is thus deprotonated on addition of acid. The process is fully reversible, i.e. on addition of sodium hydroxide, the protonated

---

[44] H. Yamauchi, Y. Takao, M. Abe, K. Ogino, *Langmuir*, **1993**, *9*, 300
[45] a) J.-H. Fuhrhop, *Angew. Chem.*, **1974**, *86*, 363; *Angew. Chem., Int. Ed. Engl.*, **1974**, *13*, 321
b) J.-H. Fuhrhop, U. Wanja, M. Bünzel, *Liebigs Ann. Chem.*, **1984**, 426

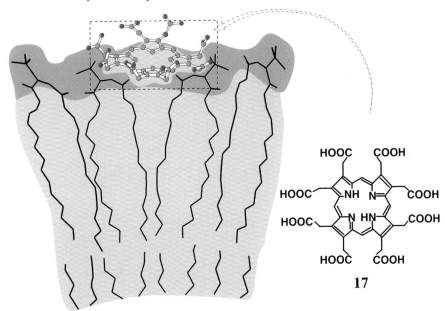

Figure 4.14 *Octaacetic acid porphyrin 17 is reversibly adsorbed on vesicle surfaces below pH 4.5. At pH 5, the porphyrin dissolves in water. It is now protonated at the pyrrolenine nitrogen atom, whereas the acetic acid side chains are half-deprotonated. At higher pH, the two-banded spectrum of the N-diprotonated porphyrin dication is again replaced by a four-banded spectrum of the neutral porphyrin base.*

porphyrin cation reappears. When the water-soluble porphyrin is localized in the inner water region 9, the spectroscopic 616 → 625 nm shift occurred with some delay as pH equilibration across the membrane requires time, e.g. 15 minutes for lecithin membranes containing 20% of cholesterol. The reversibility of the incorporation process shows that the porphyrin is dissolved peripherally in the hydrophobic region 4 (Figure 4.14). It was also discovered that the vesicle dissolved porphyrin was *not* protonated in strongly acidic media. At pH = 1 the four-banded spectrum of the free porphyrin base was still present[31]. The vesicular membrane is therefore much more "organic" or "dry" than the micellar core in respect to chemical reactivity.

The zinc complex **18** of *meso*-viologen octaethylporphyrin is more soluble in water than in chloroform. However, the moment it is mixed with vesicles, it is integrated into the membrane (Figure 4.15). The zinc ion cannot be removed by strong acids and the porphyrin cannot be oxidized to the radical by water-soluble oxidants. In electroneutral or electronegative vesicles, however, the viologen substituent is reduced to the radical by sodium dithionite. Accordingly, the porphyrin should be localized in the regions 4 and 5. In analogy with the "integral proteins", the "integral porphyrin" can only be removed from the vesicle membrane if the membrane is destroyed by detergents or solvents, otherwise it does not appear in water[45].

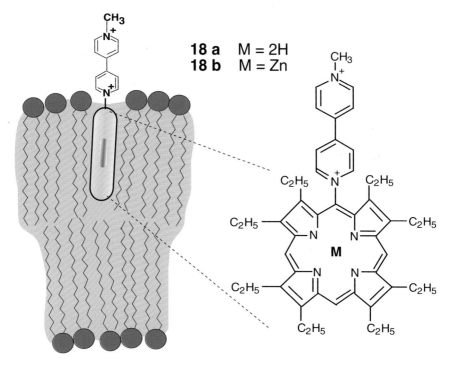

Figure 4.15 *Amphiphilic porphyrins with only one polar side can be integrated into vesicle membranes. Polar external reagents do not reach these porphyrins.*

Fluorescence quenching studies[46] showed that chlorophyll a molecules are solubilized by dimyristoylphosphatidylcholine (DMPC) only up to concentrations of around 1 mol %. Excess molecules precipitate almost exclusively in domains containing chlorophyll. Pheophytin a solubility in liquid DMPC amounts to about 10 mol %.

Flash photolysis of vesicle membranes containing chlorophyll in region 4, an entrapped water-soluble quinone (Q) in region 9 and reduced glutathione (= GSH) in the bulk phase 1 (see Figure 4.13) yielded a decay rate of 1.3×10^7 M^{-1} s^{-1} for the excited chlorophyll triplet state[47]. The resulting chlorophyll cation radical (Chl$^{·+}$) was reduced by the glutathione in the bulk phase with $k = 2.6 \times 10^6$ M^{-1} s^{-1} in competition with the recombination between Chl$^{·+}$ and Q$^{·-}$ ($k = 2.5 \times 10^3$ s^{-1}). Up to 20% of all photons absorbed by the vesicle system resulted in an **electron transfer across the membrane** from GSH to Q (Figure 4.16a).

Similar results were obtained when the zinc porphyrin was bound to the outer vesicle surface and a quinone bolaamphiphile was integrated within a DHP or DODAB vesicle membrane[48]. The quenching constant of the porphy-

---

[46] J. Luisetti, H.-J. Galla, H. Möhwald, *Ber. Bunsenges. Phys. Chem.*, **1978**, *82*, 911
[47] W.E. Ford, G.E. Tollin, *Photochem. Photobiol.*, **1983**, *38*, 441
[48] U. Siggel, H. Hungerbühler, J.-H. Fuhrhop, *J. Chim. Phys.*, **1987**, *84*, 1055

rin triplet state in the presence of the membrane bound quinone was $k = 6.2 \times 10^7$ M^{-1} s^{-1} for DODAB vesicles and the electron transfer rates in both directions exceeded 10^6 M^{-1} s^{-1} (Figure 4.16b). Although the quinone cannot be reduced by sodium borohydride, its accessibility by the photo-excited external porphyrin is easy. Negatively charged amphiphiles usually enhance the lifetime of Chl$^{\cdot+}$–Q$^{\cdot-}$ ion pairs because they eject Q$^{\cdot-}$ radicals into the water phase[49]. When a zinc *meso*-tetrapyridinium porphyrinate has bound to the

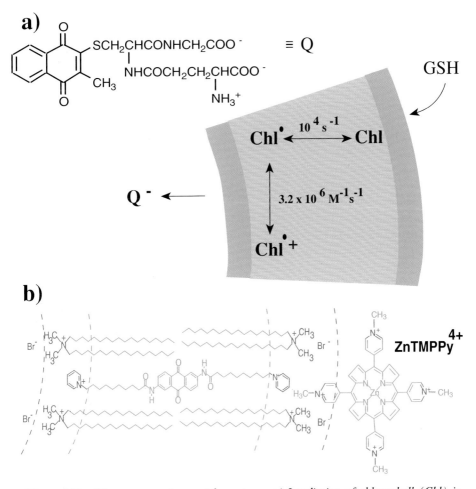

Figure 4.16 Charge separating vesicle systems. a) Irradiation of chlorophyll (Chl) in vesicle membranes leads to an efficient electron transport from GSH to entrapped quinones (Q)[47].
b) Water-soluble porphyrins photoreduce membrane-integrated quinones within microseconds. The back-reaction is equally fast[48].

[49] a) J.K. Hurst, L.Y.C. Lee, M. Grätzel, *J. Am. Chem. Soc.*, **1983**, *105*, 7048
b) J.K. Hurst, in Kinetics and Catalysis in Microheterogeneous Systems, Surfactant Science Series, Volume 38, M. Dekker, New York, **1991**, 183

surface of negatively charged vesicles, the formation of a ZnP$^{\cdot+}$–ZnP$^{\cdot-}$ ion pair with a decay rate of $t = 20$ ms was found[49]. This compares very favourably with all other membrane separated porphyrin systems reported so far. On the other hand, quenching of the porphyrin π-anion by N-methyl-N'-tetradecylbipyridinium happened very quickly with $k_Q = 1.7 \times 10^9$ M^{-1} s^{-1}.

It may therefore be concluded that systems containing Zn^{II} porphyrins in solution with membrane-bound acceptors could be expected to be more practicable than other configurations in charge separation systems which rely upon generation of vesicle membrane separated redox products.

The photoionization of zinc tetraphenylporphyrin and chlorophyll a in vesicles was also followed up by the ESR spectroscopy of rapidly frozen vesicles[50,51]. It was found that the radical cation yield decreases in negatively charged vesicles and increases in positively charged vesicles. The cationic vesicles obviously favour electron escape from the photoproduced cation via the vesicle interface into bulk water; negatively charged surfaces hinder the escape.

In connection with these facts an earlier result with covalently connected porphyrin–quinone pairs should be mentioned. In frozen solvents, it was shown that a pair linked by a diamide chain produced porphyrin–quinone radical pairs which were stable for hours at room temperature, whereas a diester analogue displayed a rapid decay of ESR signals[52]. It was also shown that the porphyrin–quinone distance must be 10–12 Å and that only 1–2% of the covalent-bound redox pair was frozen in the correct conformation giving the long-lived charge separation. It is possible that hydrogen bridged amide chains may be involved here, which are responsible for the infinite stabilization of ion pairs. This eventuality, with regard to membrane systems in which the porphyrin and electron acceptors are separated by an amide-bonded system, should also be investigated.

Various amphiphilic porphinato iron(II) complexes in vesicles were applied as oxygen-binding systems, eventually useful as **"artificial blood"** [53,54]. For example, the picket fence porphyrin **19** was embedded in a partially polymerized vesicle membrane and then 1-dodecylimidazole was added. This system reversibly bound molecular oxygen[53] at pH 6 with 40 torr half-saturation pressure. The vesicular solution could be stored for months without precipitation. 11.6 mL of gaseous O$_2$ were bound per 100 mL of solution. Low pH values stabilized the oxygen complex. At pH 8 around half of the oxygen was released. This action was explained by a deprotonation of 1-dodecylimidazole within the vesicle which converted a mono-imidazole adduct into a less favourable bis-imidazole adduct. A similar porphyrin with only one amphiphilic chain and a phosphatidyl serine head group showed a comparatively lower oxygen-binding

[50] M.P. Lanot, L. Kevan, *J. Phys. Chem.*, **1989**, *93*, 998
[51] T. Hiff, L. Kevan, *J. Phys. Chem.*, **1989**, *93*, 2069
[52] A.R. McIntosh, A. Siemiarsczuk, J.R. Bolton, M.J. Stillman, R.-F. Ho, A.C. Weedon, *J. Am. Chem. Soc.*, **1983**, *105*, 7215
[53] E. Tsuchida, N. Nishide, M. Yuasa, *J. Chem. Soc., Chem. Commun.*, **1986**, 1107
[54] E. Hasegawa, M. Fukuzumi, H. Nishide, E. Tsuchida, *Chem. Lett.*, **1990**, 123

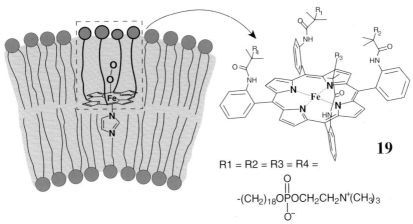

Figure 4.17 *The picket fence porphyrin reversibly adds molecular oxygen, when dissolved in vesicles containing dodecylimidazole[53,54].*

rate constant[54]. The **"hemepocket"** depicted in Figure 4.17 constructed by four bulky, amphiphilic groups is presumably of great assistance towards a fast oxygenation.

Small phosphatidylcholine vesicles ($d = 25$ nm) containing magnesium octaethylporphyrin (MgOEP) in the presence of the water-soluble electron acceptors ferricyanide or dimethylviologen produce cation radicals upon flash photolysis[55]. When the electron acceptor is present on both sides of the vesicle bilayer, approximately triple the amount of porphyrin cations are produced than by the reaction with the electron acceptor on the outside. Furthermore decay of triplet absorption is scarcely effected when MV^{2+} is primarily on the outside of the vesicles, but occurs much more rapidly where MV^{2+} is present on both sides.

These results show clearly that most of the MgOEP is located near the inner vesicle surface although much smaller than the outer surface. A similar conclusion has been drawn previously for chlorophyll a dissolved in egg phosphatidylcholine vesicles[56]. The inner part of the bilayer membrane of small vesicles is obviously a more suitable solvent or better liquid than the outer part, which is again reasonably explained by the much higher curvature of the inner part. Furthermore, at the centre of the bilayer, the same space is occupied by about half as many molecules. Several amphiphilic porphyrins with long side chains vesiculate upon sonication. The porphyrin then becomes part of the hydrophobic core[57].

Carotenoids were also integrated into bilayer vesicle membranes and were tried out as **"molecular wires"** for electron transport through the membranes. A phenylenediamine bridged dimer containing two bixin polyenes was incorpo-

---

[55] J.F. Smalley, S.W. Feldberg, S.H. Wool, *J. Phys. Chem.*, **1989**, *93*, 2570
[56] W.E. Ford, G. Tollin, *Photochem. Photobiol.*, **1982**, *36*, 647
[57] A.P.H.J. Schenning, M.C. Feiters, R.J.M. Nolte, *Tetrahedron Lett.*, **1993**, *34*, 7077

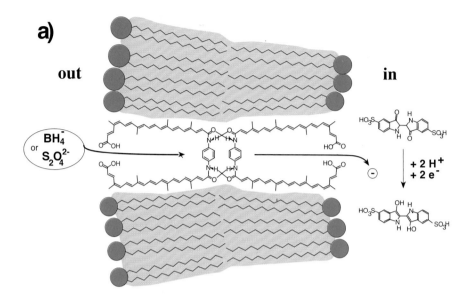

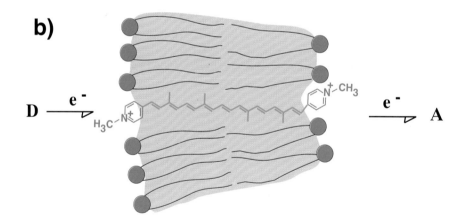

Figure 4.18 *Models for polyene wires for electron transport through vesicle membranes.
a) A phenylenediamine bridged bixin dimer facilitates the reduction of internal indigo sulfonate by external borohydride and dithionite[58].
b) Caroviologen, a polyene bolaamphiphile with pyridinium end groups, is thought to conduct electrons directly[59].*

---

[58] J.-H. Fuhrhop, M. Krull, A. Schulz, D. Möbius, *Langmuir*, **1990**, *6*, 497
[59] S.-I. Kugimiya, T. Lazrak, M. Blanchard-Desce, J.-M. Lehn, *J. Chem. Soc., Chem. Commun.*, **1991**, 1179

rated into DPPC vesicles (Figure 4.18a). An entrapped indigo dye was efficiently reduced by external dithionite and borohydride ions. However, the reduction products differed, showing that the electrons were not the only reductants. Dithionite and borohydride ions probably passed the vesicle membrane in the environment of the rigid chromophore. Methylated polyenes obviously render membranes permeable to ionic reductants[58]. Nevertheless, it was claimed that the α,ω-bipyridinium polyene ("caroviologen") with only 18 polyene carbons acts as a "wire" through a DMPC membrane containing 28 CH_2 groups (Figure 4.18b). The reduction of entrapped ferricyanide by dithionite was only enhanced by a factor of up to 8 in comparison to carotene-free DMPC vesicles. A turnover of approximately 50 electrons per wire per minute was estimated, but there was no solid proof for the "wire" character of the carotene[59]. Both "wires", the bixin derivative as well as the caroviologen, are however potent candidates for ground state electron transfer because the electron withdrawing groups should promote electron uptake in a kind of reversible one-electron Michael addition.

Finally it should not be omitted to mention the fact that the rigid sphere fullerene, C_{60}, can also be dissolved in vesicle membranes[60]. When dissolved in hexane, chloroform or 1,2-dichloroethane, a narrow, concentration-independent absorption band at 334 nm ($\epsilon \approx 52000$) was produced. In vesicles (lecithin, DODAB, DHP), the fullerene adsorption becomes concentration dependent whereby band-broadening, bathochromic shifts (343–360 nm) and loss of extinction ($\epsilon \approx 10000 \rightarrow 4000$) were observed in more concentrated solutions. C_{60} clearly aggregates within the vesicle membranes, a step not observed in micellar solutions.

4.5 Ion Transport, Domain Formation and Pores in Vesicle Membranes

Approximately 3000–4000 water molecules per second cross the phospholipid bilayer membrane of a vesicle with a head group area of ≈ 70 Å^2, but it takes 70 hours for one sodium ion. Membranes are ion-impermeable and osmotically active. These subjects have been treated in other text books[1–4] and are of no concern here; instead, we concentrate on the organic chemistry of the membrane barrier, and its strengthening, perforation and disruption by synthetic systems.

The vesicle bilayer acts as a barrier of medium strength for water-soluble, non-ionic organic compounds. Sucrose solutions, for example, were applied as contrast agents in the light microscopy of stearic acid vesicles[14]. It took hours for the carbohydrate to penetrate this barrier. Efflux rates of radioactive sucrose were measured in cationic vesicles in the presence of different counterions. For this purpose, external sucrose was removed by ultrafiltration through Nucleopore polycarbonate with 0.4, 0.2 and 0.1 μm pores. The efflux rates were then determined by dialysis methods. In the case of dihexadecyldimethyl-

[60] H. Hungerbühler, D.M. Guldi, K.-D. Asmus, *J. Am. Chem. Soc.*, **1993**, *115*, 3386

ammonium membranes, it was found that the malonate counterion in system 1 blocked the sucrose transport 40 times better than the dipalmitate system 2; only the divalent counterion managed to "patch up" leaky membranes[61]. Steroids with alkyl chains on C-17 reduce glucose leakage in DPPC vesicles, i.e. cholesterol is most effective[62].

Sucrose permeability has also been related to the stereochemistry of the head groups[63]. Three diastereomeric five-membered ring ketals were used as head groups in vesicle formation. With regards to the 1,3-dioxolane ring, the quaternary ammonium head group in **20** is *trans* to both hexadecyl chains. This arrangement produced the least sucrose-permeable vesicle membrane with the highest melting point in DSC (34.2°) and gave the most condensed monolayer with the dioxolane ring parallel to the water surface (≈ 80 Å^2/molecule). In the diastereomers like **21**, where one or both alkyl chains are *cis* to the head group, the packings produced in both vesicles and monolayers were not so tight.

Unlike conventional, one-headed amphiphiles which can readily slither into vesicle membranes as single chains, bolaamphiphiles can only insert themselves into a preformed membrane in the form of loops (Figure 4.19). Bolaamphiphiles with stiff ethylene glycol segments tend to disrupt vesicle membranes and are extremely effective in releasing entrapped dyes, e.g. 5(6)-carboxyfluorescein[64].

Bilayer vesicles, especially the non-isotonic ("stressed") kind or those which contain large amounts of dissolved cholesterol, can also be made to leak with the aid of "molecular harpoons". These are membrane-disruptive surfactants comprised of a rigid, wedge-shaped hydrophobic unit attached to a hydrophilic chain, e.g. **22**. It was discovered that just one harpoon molecule per ten

a membrane disrupting harpoon

[61] Y.-C. Chung, S.L. Regen, *Langmuir*, **1993**, *9*, 1937
[62] H. Yamauchi, Y. Takao, M. Abe, K. Ogino, *Langmuir*, **1993**, *9*, 300
[63] D.A. Jaeger, W. Subotkowski, J. Mohebalian, Y.M. Sayed, B.J. Sanyual, J. Heath, E.M. Arnett, *Langmuir*, **1991**, *7*, 1935
[64] S. Nagawa, S.L. Regen, *J. Am. Chem. Soc.*, **1991**, *113*, 7237

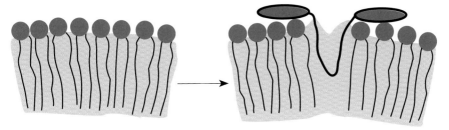

Figure 4.19 *Bolaamphiphiles destabilize bilayer vesicle membranes[64].*

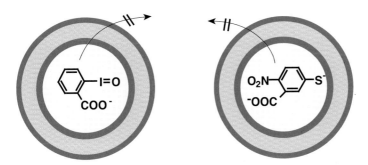

Figure 4.20 *Vesicle entrapped organic ions react slowly with each other[66].*

thousand 1-palmitoyl-ω-oleoyl-sn-glycero-3-phosphocholine (= POP) molecules has an effect on releasing vesicle entrapped fluorescent dyes[65].

Vesicles of dioctadecyldimethylammonium chloride (DODAC) are relatively resistant to permeation of organic anions at 25 °C, at imposed *exo/endo* pH gradients of 3–4 units[66]. Thiophenolate anions, for example, can *not* be oxidized by *o*-iodosobenzoate if both reagents are encapsulated in DODAC vesicles (Figure 4.20). Reaction occurs immediately after addition of some ethanol.

The entrapping of water-soluble compounds in the small internal water volume and their release within a period of several minutes or hours is evidently a relatively simple task as was demonstrated with the preceding examples. No domain formation or perforation is necessary here. Small cations, however, were only released in under a day in the case where the membrane was perforated or contained a dissolved carrier system. Before turning to these systems, we shall firstly introduce **synkinetic domains**.

Fluid, natural cell membranes are based on phospholipids which contain fatty acid esters of various chain lengths. Such lipids formed random mixtures, separated into domains or produced a mixture of both, e.g. phospholipid dimers AA and BB, in which two amphiphiles are connected by disulfide bridges. Equilibration via reduction (thiolate) – oxidation (disulfide) cycles

[65] K. Naka, A. Sadownik, S.L. Regen, *J. Am. Chem. Soc.*, **1993**, *115*, 2278
[66] R.A. Moss, S. Swarup, H. Zhang, *J. Am. Chem. Soc.*, **1988**, *110*, 2914

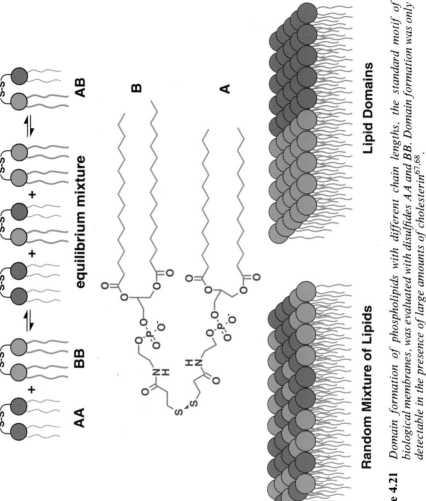

Figure 4.21 *Domain formation of phospholipids with different chain lengths, the standard motif of biological membranes, was evaluated with disulfides AA and BB. Domain formation was only detectable in the presence of large amounts of cholesterin*[67,68].

should preferentially produce AB dimers (III) in random mixtures and homodimers (I and II) if A and B form domains. With differences in chain lengths of up to four methylene groups (stearic to myristic esters) random distribution was found in the fluid and gel parts of the vesicle membrane. No domain formation had occurred. If, however, the vesicles were made more rigid by the addition of 40% of cholesterol, the random distribution of I/III/II = 1:2:1 changed to 1/1.55/1 at 60 °C. The higher melting homodimer II obviously enriches in domains and the formation of AB is partly suppressed[67,68] (Figure 4.21).

Certain pathological cells produce excessive amounts of a particular enzyme, e.g. neuroblastoma/acetylcholinesterase. If such an enzyme is capable of breaking open specially designed vesicles which encapsulate a cytotoxic drug, then cell selective therapeutic activity can be achieved. The double-chain amphiphile **23** forms vesicles and incorporates an acetylcholine-like head group which was readily hydrolyzed by acetylcholinesterase (AcE). The primary alcohol then ejected one of the two tails and the remaining single chain amphiphile destroyed the bilayer; membrane-bound fluorescent dyes were set free within seconds[69].

$$C_{16}H_{33}O-\overset{O}{\underset{}{\|}}-\overset{CH_3}{\underset{CH_3}{\overset{|}{N^+}}}-CH_2CH_2OAc \quad Br^-$$
$$C_{16}H_{33}O-\underset{O}{\overset{}{\|}}$$
23

Domain formation can also be steered from the bulk phase. A systematic organizing of parts of the membrane interior (= synkinesis of domains) was achieved by externally controlling the displacement of membrane synkinons. In mixed BLM vesicles – from (inert) dihexadecyldimethylammonium bromide and an azobenzene amphiphile – the latter compound existed as a protonated monomer at pH 4 (λ_{max} = 355 nm). At pH 7, it aggregated within the vesicle membranes and the absorbance maximum shifted to 312 nm, i.e. domains were formed. If copper sulfate is added to the solution at pH 4 the maximum was also found at 312 nm[70]. Sodium sulfate has an analogous effect, but copper chloride is ineffective. This formation and dissolution of amphiphile aggregates in a membrane is thus controlled by external protons or divalent anions or cations and constitutes a perfect example of the **steering of domain formation in which the head groups act as receptors** (Figure 4.22).

A covalent pore required for bilayers must be approximately 40 Å long and contain a polar core for ion transit and a non-polar exterior for membrane interaction. Oligopeptides with high helical content[71], polymeric crown ether[72] and "bouquet"-shaped crown ethers[73] have been reported. The only such

---

[67] S.M. Krisovitch, S.L. Regen, *J. Am. Chem. Soc.*, **1992**, *114*, 9828
[68] S.M. Krisovitch, S.L. Regen, *J. Am. Chem. Soc.*, **1993**, *115*, 1198
[69] F.M. Menger, D.E. Johnston, jr., *J. Am. Chem. Soc.*, **1991**, *113*, 5467
[70] M. Shimomura, T. Kunitake, *J. Am. Chem. Soc.*, **1982**, *104*, 1757
[71] J.D. Lear, Z.R. Wasserman, W.F. de Grado, *Science*, **1988**, *240*, 1177
[72] U.F. Kragton, M.F.M. Roks, R.J.M. Nolte, *J. Chem. Soc., Chem. Commun.*, **1985**, 1275
[73] M.J. Pregel, L. Jullien, J.-M. Lehn, *Angew. Chem.*, **1992**, *104*, 1695; *Angew. Chem., Int. Ed. Engl.*, **1992**, *31*, 1637

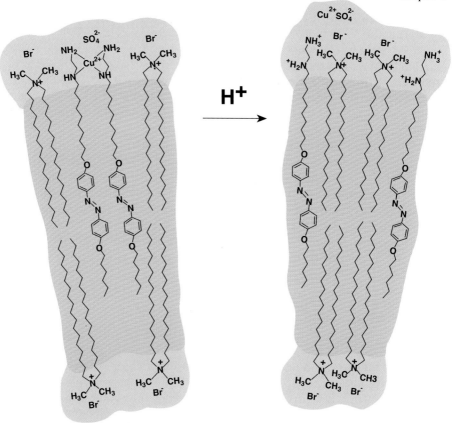

Figure 4.22 *Domain formation of amphiphiles with ethylenediamine head groups by copper sulfate complexation was detected by short wavelength shifts of azobenzene chromophores[70].*

molecule where cation transport and most importantly its inhibition were demonstrated is a conjugate of three macrocycles containing oligoethylene glycols[74]. The central unit is formed from a hexacarboxylate crown ether (Figure 4.23), the wall units prepared from six macrocyclic tetraester molecules derived from maleic acid, 1,8-octanediol and triethylene glycol. The head groups are on both sides of the bolaamphiphile, 1-mercapto-β-D-glucose. The central ring (together with the six head groups) fixates the pore within the membrane and the six ethylene glycol units provide a hydrated pore which was characterized by an elegant cation–proton transport inhibition experiment. Firstly, it was established that a cholesterol containing vesicle membrane held a stable pH gradient for times in excess of 8 h. A proton carrier (FCCP = carbonyl cyanide 4-(trifluoromethoxy)phenylhydrazone) capable of releasing protons of the acids entrapped in region 9 was then added (Figure 4.13). A

---

[74] T.M. Fyles, T.D. James, K.C. Kaye, *J. Am. Chem. Soc.*, **1993**, *115*, 12315

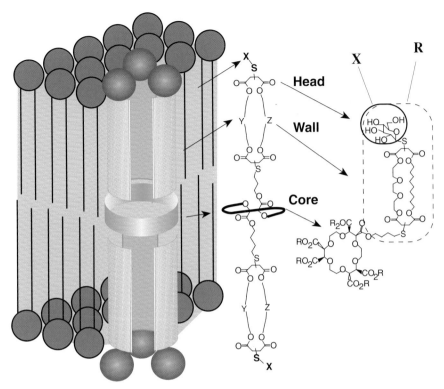

Figure 4.23 *A covalent pore for bilayer vesicle membranes*[74].

surplus of anions, however, appeared in region 9, allowing the protons to flow out of the vesicle, *only* if other cations (in this case K^+) are transported through the membrane into the vesicles. Only in the presence of both K_2SO_4 and the membrane pore, the pH in the bulk volume dropped and could be determined kinetically by titration. The ion flow could also be slowed down by lithium ions; reversible blocking of the pore by large organic cations has, however, not yet been demonstrated.

The decisive question in pore or tunnel construction is, however, whether reversible opening and closing can be achieved. Only if this can be realized by external signal molecules, can the "ion channel" become part of a machinery. MLM vesicles are particularly easy to perforate, because the membranes and pores can be as thin as 2 nm. The first pore designed for such a monolayered membrane was the natural edge amphiphile and ionophore monensin, which was converted to a tetraanionic bolaamphiphile by esterification with pyromellitic acid (Figure 4.24). 50 molecules per vesicle were sufficient enough to open the vesicle to the flow of lithium ions. Presumably between four to six monensin bolaamphiphiles formed a hydrated domain by aggregation of the ether edges with the hydrophobic membrane. The pore was sealed irreversibly via sonication in the presence of a bis-tetraalkylammonium

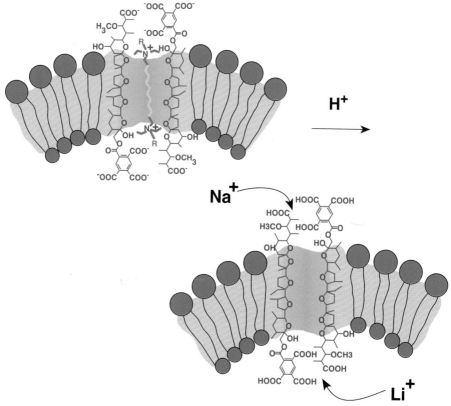

Figure 4.24 *A monensin-based, dianionic bolaamphiphile assembles to form hydrated ion pores in monolayered vesicle membranes. They can be reversibly closed by α,ω-diamino bolaamphiphiles[75].*

bolaamphiphile, but the corresponding tertiary amine was released from the pore at pH 3, when the carboxylic acid groups were protonated[75].

Upon sonication, an α,ω-disulfone macrocycle (Figure 4.25) produces extremely long-lived, electroneutral vesicles which were perforated with various PEG derivatives, but these synthetic pores could not be efficiently sealed. An oligoamino pore with anionic head groups turned out to be more controllable. It allowed vesicle entrapped calcein to be fully accessible to Fe^{II} ions which quenched its fluorescence. Taurine and several other bulky anions closed the amino pore completely and pH-reversibly[76]. It was also possible to suck the entrapped iron ions out of the vesicle with external EDTA (Figure 4.25). So far, the synkinetic pores have proved to be much more controllable than the covalent ones.

[75] J.-H. Fuhrhop, U. Liman, H.H. David, *Angew. Chem.*, **1985**, *97*, 337; *Angew. Chem., Int. Ed. Engl.*, **1985**, *24*, 339
[76] J.-H. Fuhrhop, U. Liman, V. Koesling, *J. Am. Chem. Soc.*, **1988**, *110*, 6840

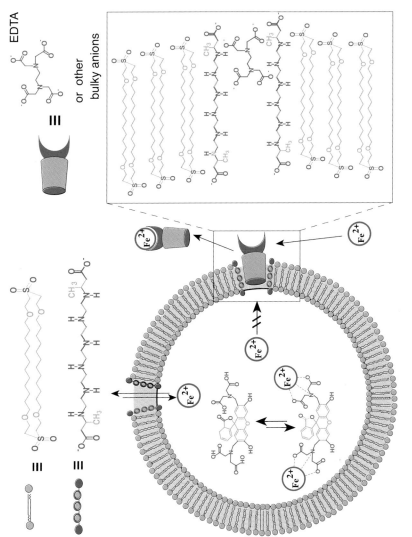

Figure 4.25 The synkinetic system of a disulfone bolaamphiphile MLM, an oligoamino-α,ω-dicarboxylate pore and an entrapped dye (calcein) can be reversibly opened and closed to iron-ion transport with EDTA stoppers[76].

The ideal pore should be opened by monochromatic light of a given wavelength and closed by light of a different wavelength. As yet such a pore has neither been obtained by covalent synthesis, nor by non-covalent synkinesis, but the following transport system comes close to it. Upon UV irradiation in apolar solvents the photospiran in Figure 4.26 is converted to the purple, ring-opened form. This zwitterionic dye forms an ionic complex with zwitterionic amino acids (e.g. Phe) or their cationic methyl esters (e.g. PheOMe). This means that if a hydrophobic amino acid is dissolved in water and photospiran in an organic phase, **the amino acid is transferred to the organic phase upon UV irradiation** as, for example, a molecular complex of PheOMe and the zwitterionic dye. If the system is dissolved in vesicle membranes, it transports amino acids through the membranes upon irradiation with UV light. The complex between the two zwitterions diffuses relatively rapidly through the membrane. Twenty minutes of UV irradiation were then followed by 5 min of visible light in order to regenerate membrane-soluble spiran, enabling the membrane interior to once more receive an amount of transport agent. Further employment of UV irradiation ensured the continuation of the amino acid transport. Claims have even been staked that outer-surface-adsorbed PheOMe was transported against a concentration gradient into the vesicle[77]. Photochromism often becomes irreversible in surface monolayers, where zwitterions stabilize in the form of 2D crystals[78].

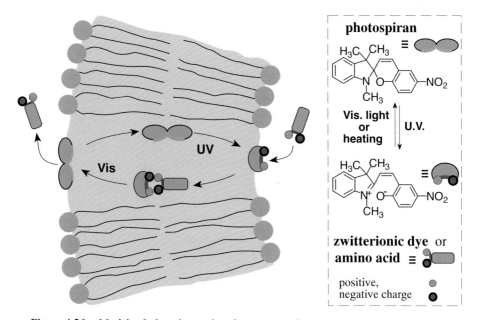

Figure 4.26 *Model of the photoinduced amino acid transport through vesicle membranes[77].*

[77] J. Sunamoto, K. Iwamoto, Y. Mohri, T. Kominato, *J. Am. Chem. Soc.*, **1982**, *104*, 5502
[78] E. Ando, K. Moriyama, K. Arita, K.A. Morimato, *Langmuir*, **1990**, *6*, 1451

4.6 The Entrapped Water Volume

Surfactant vesicles entrap electrolytes and polyelectrolytes. Only protons and hydroxide ions move freely across the mono- and bilayers of vesicles. However, in vesicles which contain cholesterol a pH gradient may stay stable for up to an hour[32] (see also page 80 and Figure 4.23).

Vesicles with entrapped metal ions can be separated from metal ions in the bulk phase by gel chromatography[11,33]. This is a tedious procedure and can be avoided through the employment of high valent cations which can be regioselectively masked with charged water-soluble ligands. A number of paramagnetic lanthanide anions, e.g. the dysprosium nitrilotriacetate complex $Dy(NTA)_2^{3-}$, act as shift reagents for such metal ions ($^{23}Na^+$, $^{39}K^+$, $^{43}Ca^{2+}$) in aqueous media. The point of membrane transport studies is usually to ensure that only the shift reagent present inside closed vesicles is acquired. The easiest way to accomplish this is firstly to prepare the vesicles in the presence of the shift reagent anions and secondly, to selectively deactivate purely the outside shift reagent; a simple achievement via titration with the diamagnetic Lu^{3+} ion. Equimolar amounts of Lu^{3+} deactivate the $Dy(NTA)_2^{3-}$ anions by the formation of electroneutral DyNTA and LuNTA complexes[79-81]. It has also been demonstrated that an increase in charge of the complex anion produces a more efficient shift reagent, due to the increase of the fraction of ligand bound metal ions.

Quantitative entrapment of DNA[82] and negatively charged colloids[83] could be achieved through bolaamphiphiles with one electropositive and one electroneutral head group. The ammonium head binds as counterion to the polyelectrolyte; the hydrophobic effect enforces a closed membrane structure and the carbohydrate head group assures water-solubililty (Figure 4.27). This procedure is the easiest way to synkinetize redox-active systems with membrane dissolved organic dyes and inorganic oxidants for charge separation experiments.

A number of water-soluble dyes, e.g. calcein, have also been entrapped and can be used for the detection of ion transport (see Figure 4.25).

4.7 Polymerization in Vesicle Membranes and Interactions with Polymers

Biomembranes contain up to 50% of "membrane proteins". Their chains usually fold to form regular groups of α-helices in the part of the protein which traverses the hydrophobic core of the bilayer. The helices are often covalently

---

[79] M.M. Pike, C.S. Springer jr., *J. Magn. Reson.*, **1982**, *46*, 348
[80] S.C. Chu, M.M. Pike, E.T. Fossel, T.W. Smith, J.A. Balschi, C.S. Springer jr., *J. Magn. Reson.*, **1984**, *56*, 33
[81] M.M. Pike, D.M. Yarmush, J.A. Balschi, R.E. Lenkinski, C.S. Springer jr., *Inorg. Chem.*, **1983**, *22*, 2388
[82] J.-H. Fuhrhop, H. Tank, *Chem. Phys. Lipids*, **1987**, *43*, 193
[83] B. Henne, Dissertation Freie Universität Berlin, **1992**

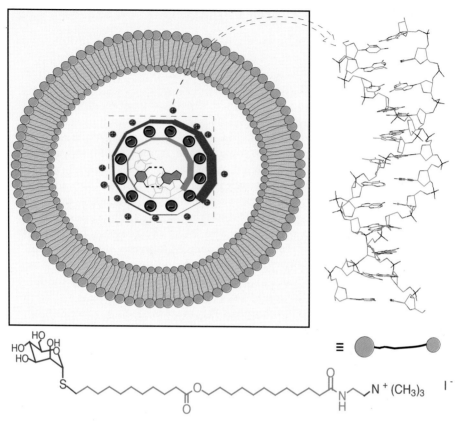

Figure 4.27 *Long-chain bolaamphiphiles with an electroneutral and a cationic head group form membranes around anionic polymers (here: DNA) and colloids. Synkinesis of organic–inorganic composite materials in bulk aqueous media can thus be achieved[82].*

connected by turns and bends of the protein backbone which are located outside the membrane, but the helical units may also be parts of different molecules in a protein cluster. Many membrane proteins have been extracted from biomembranes with detergents, e.g. octylglucoside; some were crystallized and reconstituted with artificial vesicle membranes. The hydrophobic outer part of the helical protein is then tightly integrated into the vesicle membrane and no ions flow at the protein–membrane interface[84].

What happens if the lipid molecules of an artificial membrane themselves contain polymerizable groups and are polymerized after a vesicle membrane has been formed from the monomers? Will the polymerization chain reaction run through the whole of a monolayer and will the polymer retain the vesicle structure? Or will parallel ordered clusters be formed and will the vesicle be ruptured? The answers to most of these questions are frustrating: domains do

---

[84] A.H. Parola, in Biomembranes (M. Shinitzky, ed), VCH, Weinheim, **1993**, p. 189ff

appear and the vesicle is highly disturbed by polymerization, if the outer head groups are connected.

Most studies with polymeric vesicles have been carried out with vesicles made of monomeric, polymerizable amphiphiles. In respect to stable, biodegradable vesicles, possibly applicable as drug carriers, the polycondensation of L-amino acid head groups is of particular interest. Back in 1948, the first studies on the condensation of octadecyl esters of alanine in LB multilayers were published. Here, the esters spontaneously aminolysed and produced water-soluble polypeptides[85]. Later polycondensated vesicles were prepared from maleic anhydride, dioctadecylamine and cysteine. To induce polycondensation, N-cyclohexyl-N'-[β-(N-methylmorpholino)ethyl]carbodiimide-p-toluenesulfonate was introduced as the water-soluble condensing agent[86]. Apart from the polyamide formation several side-reactions occurred, but well-defined liposomes remained in solution. The behaviour of monomeric versus polymeric liposomes towards Triton-X-100, however, did not change. A real stabilization of the vesicle membrane could not be achieved and the average degree of polymerization was found to be only in the order of 4. Similar results were previously obtained from other head group polymerizations.

Prepolymerized lipids form vesicles only if the disentanglement of the polymer main chain (= back bone) and the membrane forming side-chains is simplified by a hydrophilic spacer between them[87,88]. Efficient decouplings of the motions of the polymeric chain and the polymeric bilayer are thus achieved and stable liposomes with diameters of around 500 nm were formed upon ultrasonication (Figure 4.28a). Their bilayer showed a well-defined melting behaviour in DSC. The ionene polymer with C_{12}, C_{16} and C_{20} intermediate chains also produced vesicles upon sonication (Figure 4.28b). Here, the amphiphilic main chain is obviously so simple that ordering to form membranes produces no problems whatsoever[87].

Another possibility for obtaining polymerized vesicles is through the linkage of monomers by disulfide groups. 1,2-bis(11-mercaptoundecanoyl)-sn-glycero-3-phosphocholine, for example, can be used to construct vesicles which can be oxidatively polymerized using hydrogen peroxide as well as being reductively depolymerized via dithiothreitol[89]. The number average degree of polymerization was estimated to be 25. The question as to whether or not coupling occurs across the bilayer was answered by freeze fracture and electron microscopy. The freeze fracture technique is known to generate two complementary fracture faces of a bilayer membrane, a result of the splitting of the bilayer into two monolayers. If, however, both monolayers are connected by covalent bonds the fracture plane will no longer run through the middle of the bilayer, the vesicle disrupts completely, leaving rings of polymeric debris. Polymerized

---

[85] A. Baniel, M. Frankel, I. Friedrich, A. Katchalsky, *J. Org. Chem.*, **1948**, *13*, 791
[86] R. Neumann, H. Ringsdorf, *J. Am. Chem. Soc.*, **1986**, *108*, 487
[87] T. Kunitake, N. Nakashima, K. Takarabe, M. Nagai, A. Tsuge, H. Yanagi, *J. Am. Chem. Soc.*, **1981**, *103*, 5945
[88] R. Elbert, A. Laschewsky, H. Ringsdorf, *J. Am. Chem. Soc.*, **1985**, *107*, 4134
[89] N.K.P. Samuel, M. Singh, K. Yamaguchi, S.L. Regen, *J. Am. Chem. Soc.*, **1985**, *107*, 42

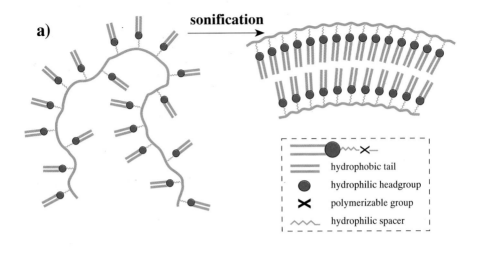

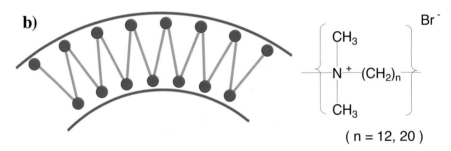

Figure 4.28 *a) Disentanglement of the polymer main chain from the membrane surface by hydrophilic spacers, often ethylene glycol oligomers, is the best means to synkinetize vesicles from amphiphilic polymers. Ordering of the amphiphiles must not be disturbed by their covalent connections[84].
b) Folding of the ionene polymer spontaneously yields monolayer vesicle membranes[87].*

samples only produced three-dimensional, half-balls in freeze-fracture micrographs, which suggests polymerization within the monolayers[90]. Cross-linked bilayers are "cross-fractured", generating the morphology of circles instead of spheres (Figure 4.29). The degree of polymerization was again found to be about 25 and the vesicle stable to ethanol.

The best vesicle stabilization effect was obtained by polymerization of butadiyne units within the hydrocarbon chains of amphiphiles after the vesicle was formed. This polymerization produces red or blue polyenes and occurs only if the vesicle membrane is in the liquid crystalline state. No polymers formed

---

[90] M.F. Roks, H.G.J. Visser, J.W. Zwikker, A.J. Verkley, R.J.M. Nolte, *J. Am. Chem. Soc.*, **1983**, *105*, 4507

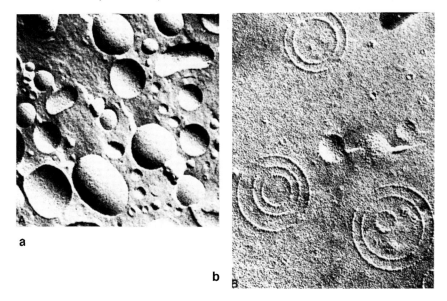

Figure 4.29 *Freeze fracture electron micrographs of a) unpolymerized and b) polymerized vesicles[90]. Magnification 6.8×10^4.*

above the phase transition temperature. Bilayers with oligoacetylene units are not only well integrated into vesicle membranes which do not release entrapped ions, but they are also stable to 50% ethanol. Since diyne amphiphiles and bolaamphiphiles are easy to synthesize (see Scheme 2.14), they have become the standard for polymerized lipids. Polymerized vesicles, formed through irradiation of a vesicle membrane containing 90% of the diyne and 10% of the non-polymerizable lipid, were transformed to stable vesicles with large holes ("2D sponges") after treatment with ethanol[91]. The polymers formed a coherent net of small oligomers which were insoluble in ethanol, whereas non-polymerizable lipids also occurred in domains which could be removed. Giant vesicles with holes ("skeletonized vesicles") were also seen through the light microscope[92].

Polymerized lipids do not occur in natural cell membranes. Nature tends to support fragile membrane structures with polymeric skeletons, i.e. protein cytoskeletons, polysaccharide cell walls etc. Analogous synthetic polymeric nets are simply constructed from polymerizable counterions. Negatively charged dihexadecyl phosphate vesicles can be neutralized with choline methacrylate; polymerization of the latter produces a polycationic vesicle coat which is *not* inserted into the membrane (Figure 4.30)[93]. A cytoskeleton at the

---

[91] E. Sackmann, P. Eggl, C. Fahn, H. Bader, H. Ringsdorf, M. Schollmeier, *Ber. Bunsenges. Phys. Chem.*, **1985**, *89*, 1198

[92] H. Ohno, S. Takeoka, E. Tsuchida, *Polymer Bull.*, **1985**, *14*, 487

[93] H. Ringsdorf, B. Schlarb, J. Venzmer, *Angew. Chem.*, **1988**, *100*, 117; *Angew. Chem., Int. Ed. Engl.*, **1988**, *27*, 113

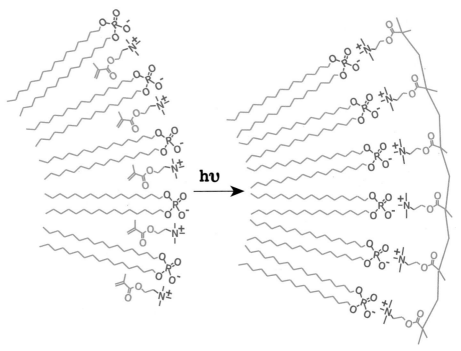

Figure 4.30 *Light-induced polymerization of cationic choline acrylate counterions on negatively charged vesicle surfaces produces a polymer net around the vesicle*[93].

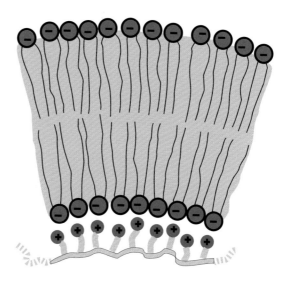

Figure 4.31 *The same procedure as in Figure 4.30 with entrapped choline acrylate gives an internal net, or a "cyto-skeleton"*[93].

inner surface of vesicles can be obtained analogously: liposomes made of 90 mol % phosphate diester and 10% of methacrylate were mixed with choline methacrylates and sonicated. Gel chromatography removed the outside methacrylate molecules and UV irradiation gave an inner membrane-bound, polycationic coating (Figure 4.31). Such coated vesicles show the same permeability to [^{3}H]-glucose as the non-coated precursors[94].

Molecular interactions between D-glucolipids and multilayered vesicles were also demonstrated with partially polymerized diacetylene units. Red vesicles constructed by sonication and subsequent UV irradiation formed a red precipitate with Con A which was redissolved by methyl-D-mannopyranoside to regenerate a red solution[95,96].

External polymers can also be employed to control leakage of entrapped solutes. Adsorption of the hydrophobic polyelectrolyte poly(2-ethylacrylic acid) on phosphatidylcholine membranes causes the disruption of vesicle membranes (Figure 4.32) with small changes in pH (7.4 → 6.5)[97]. The surface polyelectrolyte undergoes a well-defined conformational transition from an

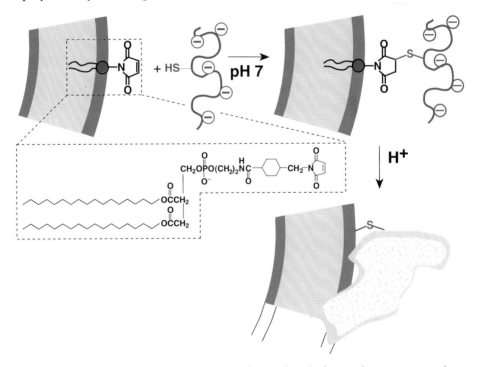

Figure 4.32 *Polyanions on vesicle surfaces do not disturb the membrane structure, but globular polymers disrupt the membrane[97].*

[94] H. Ringsdorf, B. Schlarb, P.N. Tyminski, D.F. O'Brien, *Macromolecules*, **1988**, *21*, 671
[95] L. Gros, H. Ringsdorf, H. Schupp, *Angew. Chem.*, **1981**, *93*, 311; *Angew. Chem., Int. Ed. Engl.*, **1981**, *20*, 305
[96] H. Bader, H. Ringsdorf, J. Sikura, *Angew. Chem.*, **1981**, *93*, 109; *Angew. Chem., Int. Ed. Engl.*, **1981**, *20*, 91
[97] M. Maeda, A. Kumano, D.A. Tirrell, *J. Am. Chem. Soc.*, **1988**, *110*, 7455

expanded, hydrophilic coil to a compact globular structure. The globular form binds strongly to the vesicle surface and renders the membrane permeable. The effect is similar to the loosening of the vesicle membrane by the "flippase" model (see Figure 4.8). Photoinduced acidification has also been used for this process[98]. For the detection of a membrane rupture, the releasing of entrapped calcein was employed in both cases[97].

4.8 Photoreactions in Vesicle Membranes

Light-induced charge separation was treated in the porphyrin section 4.4, activities of entrapped dyes in section 4.5. In the following, we discuss the effect of vesicle viscosity on photoreactions and photochemical labelling.

Vesicles are "ordered" fluids or liquid crystals, a fact which reflects well in those photoreactivities that are particular to vesicle solutions. Fatty acid derivatives **24** and **25**, for example, show low quantum yields of fluorescence (ϕ_f) and high quantum yields of *cis–trans* isomerization (ϕ_c) as well as short fluorescence life times (t_f) in both methylcyclohexane and micellar SDS solutions (Table 1). In DPPC vesicles, on the other hand, *cis–trans* isomerizations are cumbersome and much slower, and the fluorescence yield and lifetime rise considerably (Table 1). For those stilbene derivatives which are "embedded" in the middle of a fatty acid backbone, isomerization is virtually eliminated in the low-temperature or gel phase of the bilayer[99]. The vesicle thus plays the role of stabilizing *trans* configurations which fit into the frozen oligomethylene chain matrix.

$$\text{Ph-CH=CH-C}_6\text{H}_4\text{-(CH}_2)_4\text{COOH}$$
24

$$\text{CH}_3(\text{CH}_2)_3\text{-C}_6\text{H}_4\text{-CH=CH-C}_6\text{H}_4\text{-(CH}_2)_5\text{COOH}$$
25

Azobenzene and its derivatives are photoisomerized from the *trans* to the *cis* form by UV light (313 nm). Heat or visible light (420 nm) regenerate the *trans* isomer and this isomerization changes the molecular form, the UV/VIS spectrum and the dipole moment of the chromophore[100]. The *trans* form in vesicle membranes is also particularly stable below its transition temperature.

Photoaddition reactions of entrapped bolaamphiphiles should occur regioselectively if both head groups are locked on the inner and outer surface of a vesicle membrane. A detailed study of such a reaction confirmed this pre-

---

[98] H. You, D.A. Tirrell, *J. Am. Chem. Soc.*, **1991**, *113*, 4022
[99] D.G. Whitten, *Acc. Chem. Res.*, **1993**, *26*, 502
[100] H. Menzel, *Nachr. Chem. Tech. Lab.*, **1991**, *39*, 636

Table 1 Photochemical quantum yields of fluorescence (ϕ_f), cis–trans isomerization (ϕ_c) and fluorescence lifetimes (τ_f) for stilbene-fatty acid derivatives in free solution, micelles and vesicles.

Stilbene	Medium	ϕ_f	ϕ_c	τ_f
24	methylcyclohexane	0.08	0.48	210
24	SDS micelles	0.12	0.44	260
24	DPPC vesicles	0.51	0.15	1400
25	methylcyclohexane	0.21	0.37	500
25	DPPC vesicles	0.92	0.01	1780

diction although the bolaamphiphile was not properly oriented in the membrane[101]. The conformation of the non-cyclic bolaamphiphile DIPEP (= bisphosphatidylethanolamine (trifluoromethyl)phenyldiazirine) in various vesicle membranes was probed by the reaction with two water-soluble, highly electrophilic sulfonic acids, namely N-[(m-sulfobenzoyl)oxy]sulfosuccinimide (SSSB) and trinitrobenzenesulfonic acid (TBS). Both reacted readily with the terminal amino groups of DIPEP which were located on the outer vesicle surface. Analysis of the products, which carried either one or two end group labels, indicated a population close to the statistical proportion 1:2:1 for outwardly U-shaped, transmembraneous and inwardly U-shaped DIPEP conformations (Figure 4.33). Single chain bolaamphiphiles are therefore not really suitable for fixating reactive groups within vesicle membranes. Nevertheless several integral membrane proteins were then successfully photolabelled with DIPEP, whereas extramembraneous peptides showed insignificant reaction with DIPEP[101].

Vectorial photoinduced electron transfer (PET) from the vesicle dissolved triphenylbenzylborate anion to a water-dissolved, tricationic cyanine dye was related to the simultaneous binding of donor and acceptor molecules at close, but non-overlapping binding sites[102,103]. PET did not occur with the tetraphenylborate anion (Figure 4.34). In the first case PET produces a boranyl radical which may decompose to form triphenylboron and a benzyl radical. In the case of the tetraphenylborate, an unstable phenyl radical would be formed, and back electron transfer is a much more favourable process than bond cleavage. Moreover, the fluorescence of the dye is quenched by Ph_3BnB^- but not by Ph_4B^- when bound to vesicles. This observation also indicates that the irreversible bleaching of the dye is caused by electron transfer quenching and an inhibition of the back electron transfer by decomposition of the electron donor.

4.9 Rearrangements of Vesicles

One reversible transformation between vesicles and micelles via pH changes has already been discussed in the micelle chapter, namely the conversion of soap

---

[101] J.M. Delfino, S.L. Schreiber, F.M. Richards, *J. Am. Chem. Soc.*, **1993**, *115*, 3458
[102] B. Armitage, D.F. O'Brien, *J. Am. Chem. Soc.*, **1992**, *114*, 7396
[103] B. Armitage, J. Retterer, D.F. O'Brien, *J. Am. Chem. Soc.*, **1993**, *115*, 10786

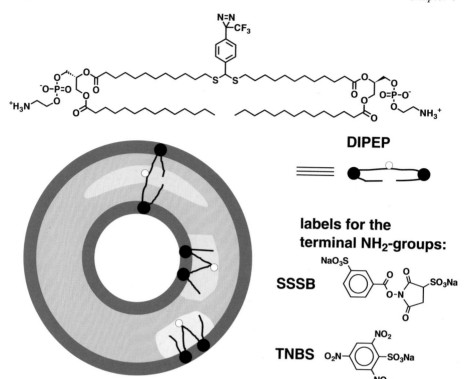

Figure 4.33 *Photolabelling experiment of a vesicle bilayer membrane with bolaamphiphile DIPEP. The light green areas indicate the membrane portions which would be photolabelled by the diazirine (white circles) upon UV irradiation, if the bolaamphiphile stretched through the membrane was inwardly or outwardly U-shaped. Labelling experiments of the amino groups on the outer surface shows that all three conformations occurred in a statistical ratio*[101].

micelles to fatty acid vesicles (section 3.9). Several similar examples have already been handled but here is one more example. The cholesterol oligo-ethylene glycol derivative **26** with a terminal carboxyl group forms bimolecular lamellae in thermotropic and lyotropic liquid crystals as well as in concentrated aqueous suspensions. Upon dilution at pH < 7, spherical vesicles are formed which dissociate to form micelles at pH ≥ 9[93,104]. The cmc is 1.6×10^{-3} M at pH 9.3. In this case, slender, flexible head groups and tails are connected to a rigid, bulky hydrophobic steroid segment. The structure of supramolecular assemblies, if any, is unpredictable for such a molecule. Under varying conditions, a variety of molecular assemblies can usually be found.

Small bilayer vesicles usually fuse to form larger vesicles and finally lamellae. Fusion of bilayer vesicles begins with a "flattening" against each other, a

---

[104] K.R. Patel, M.P. Li, J.R. Schu, J.D. Baldeschwieler, *Biochim. Biophys. Acta*, **1985**, *814*, 256; **1984**, *797*, 20

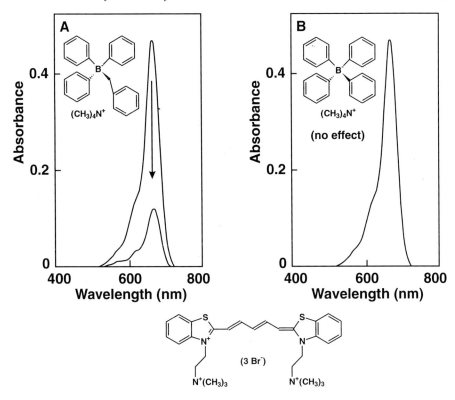

Figure 4.34 *Effect of 10 min of red light irradiation on the absorption of the water soluble, tricationic dicarbocyanine dye (λ_{max} = 669 nm) bound to negatively charged vesicle surfaces in the presence of membrane-dissolved benzyltriphenylborate (A) or tetraphenylborate (B). Only the benzyl derivative donates electrons to the dye and thereby destroys it.*

process which can be caused by the combination of a fusogenic agent (SO_4^{2-}, Ca^{2+} etc.) with the head group in the two opposed bilayers. Flattening decreases the curvature, causes dehydration of the bilayers and results in deformation and destabilization of vesicles resulting in the formation of lipidic intramembraneous particles, e.g. reversed micelles. Intermingling of the two aqueous compartments of both vesicles finally leads to a merging of adjacent bilayers which enlarges spherical vesicles[105]. The whole process of fusion may

[105] D. Yoger, B.C.R. Guillaume, J.H. Fendler, *Langmuir*, **1991**, *7*, 623

then be repeated until curvature disappears and large, lamellar structures are formed.

This process may also be reversed to enforce "budding" of large vesicles. Giant vesicles (10–200 μm in diameter) were formed simply by soaking a thin didodecyldimethylammonium bromide film in 50°C water for about an hour and then shaking it for 4 s. A light microscopic magnification of 320× was then applied to observe the giant vesicles ejecting smaller vesicles upon addition of sodium acetate solution[106]. This was explained through an exchange of bromide by an excess of acetate ions. Since the acetate only loosely binds to cationic surfaces, the outer leaflet of the bilayer membrane becomes more highly dissociated. Owing to the resulting headgroup–headgroup repulsion, the outer leaflet expands in relation to the inner one. Curvature is thus increased and budding is promoted. After approximately 15 minutes the small vesicles fuse again to giant vesicles presumably due to the small vesicles being loosely packed and migration of acetate ions restoring their equal distribution on the two sides of the vesicle walls. Vesicles stiffened with 20% cholesterol fuse very slowly.

The most important transformation of vesicles, however, is their spontaneous growth where forming tubes. Sometimes large DODAB vesicles with a queue (Figure 4.35) can be observed under the light microscope. A membrane which is curved in one dimension is directly connected here to a membrane curved in two dimensions. Thereafter, observations show that the vesicle shrinks and the queue grows[107,108], thus constituting the most simple and

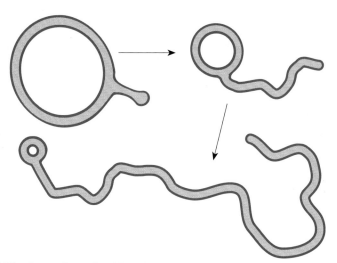

Figure 4.35 *Large dioctadecyldimethylammonium bromide vesicles may spontaneously rearrange to form tubules in aqueous suspensions[108]. Similar tubules are also observed when lipid 3D crystals swell in water (see Figure 5.1).*

[106] F.M. Menger, N. Balachander, *J. Am. Chem. Soc.*, **1992**, *114*, 5863
[107] N. Nakashima, S. Asakuma, T. Kunitake, H. Hotani, *Chem. Lett.*, **1984**, 227
[108] T. Kunitake, *Angew. Chem.*, **1992**, *104*, 692; *Angew. Chem., Int. Ed. Engl.*, **1992**, *31*, 709

fundamental rearrangement on the way from fluid spheres to fluid planar bilayers. The tubules are halfway between the vesicles and the myelin figures, the structure remaining fluid upon loss of curvature as no binding interaction is possible between tetraalkylammonium head groups in an aqueous environment. Solid 3D crystals are only viable upon dehydration.

Similar equilibria have been observed in molecular assemblies of single chain amphiphiles with a tetraalkylammonium head group, followed by a tetramethylene chain, a rigid aromatic segment and a terminal dodecyl chain. Here the supramolecular structures depend on the rigid segment structure: linear segments preferably induced the formation of vesicles; bent segments produced threads or tubules[108,109]. Mixtures of these amphiphiles supplied interesting alloys or mixed populations of supramolecular structures. Asymmetric di-n-alkyl phosphates which possess two different alcohols easily form vesicles if the chain lengths are not too dissimilar. Should the difference of chain lengths increase, e.g. C14 and C10, vesicle bilayer stability is reduced and the possibility for fusion to form crystalline tubules is favourable[110].

[109] T. Kunitake, Y. Okahata, M. Shimomura, S. Yasunami, K. Takarabe, *J. Am. Chem. Soc.*, **1981**, *103*, 5401
[110] L. Streefland, A. Wagenaar, D. Hoekstra, J.B.F.N. Engberts, *Langmuir*, **1993**, *9*, 219

CHAPTER 5

Binding Interactions in Micellar and Vesicular Fibres

5.1 Introduction

The interactions between the head groups of micelles or vesicles have so far only been related to repulsive hydration and steric forces[1]. These forces have no directional organizational power and the head groups should have rotational freedom. Iodine-laser temperature jump measurements[2] of kinetic processes in the phase transition of phospholipid bilayers showed that vesicle head groups in their liquid crystalline state are slightly fixed (with regards to the hexagonal lattice of the hydrocarbon chains), but free rotation begins at T_m with a relaxation time of around 300 ms. As yet, supramolecular organization of the monomers does *not* occur in spherical assemblies, although an intriguing earlier report exists[3] where a strong temperature-dependent CD spectrum in vesicular solutions of a double-chain glutamic acid amphiphile with a central azobenzene unit proves the contrary. Later, however, similar glutamic acid derivatives formed helical fibres in water, often growing out of vesicular structures, so it could be possible that the observed CD effect originated from some fibrous impurities rather than chiral domains on the vesicle surface.

If synkinons containing secondary amide groups are used cylinder formation is often observed upon cooling. During the transfer from a sphere to a cylinder curvature disappears in one direction. The long axis of the cylinder can be chemically stabilized by covalent polymerization, the formation of hydrogen bond chains, or other binding interactions. Repulsive hydration or steric forces are now only required in order to keep the head groups apart in the circular plane perpendicular to the cylindrical axis. The cylinders remain micellar or vesicular bilayers where all the molecules are exposed to solvent, but **their shape and organization is no longer determined by featureless repulsion or hydrophobic effects.** A most important point to be mentioned is the fact that the binding interaction between the head groups actually survives dehydration

---

[1] G. Cevc, D. Marsh, Phospholipid Bilayers, Wiley, New York, **1987**
[2] J.F. Holzwarth, *Faraday Discuss. Chem. Soc.*, **1986**, *81*, 353
[3] T. Kunitake, N. Nakashima, M. Shimomura, Y. Okahata, K. Kano, T. Ogawa, *J. Am. Chem. Soc.*, **1980**, *102*, 6642

of the fibrous assemblies. **Micellar and vesicular fibres can therefore often be lyophilized and stored as dry materials**, whereas micelles and vesicles degrade to featureless powders upon dehydration.

5.2 Fluid Molecular Bilayer Scrolls

The best-known examples of fluid rod structures are those formed by swelling of lecithin multilayer crystallites in water. They are of light-microscopic dimensions and belong to the Myelin figures, which are amphiphilic multilayer structures. During the first step, simple rods of 20–40 μm diameter (~ 250 molecular bilayers) grow into the medium and produce many irregular foldings without changing the width (Figure 5.1). The second step is characterized by much slower growth rates and the appearance of several helical and coiling forms. In the third and final step, contacting helices and coils fuse into oily streaks and complex mosaic structures[4]. The fluid character of all these structures becomes apparent through the continuous changes of the structures upon the flow of new material from the lecithin crystals to the 20–40 μm rods.

The same swelling process as in water-insoluble amphiphiles, e.g. lecithin and synthetic analogues, which produce rods, may also lead to water-filled tubular vesicles (Figure 5.2). The surface of these closed tubular structures of light-microscopic dimensions is again of a fluid character, where the size and shape of the tubules constantly change[5,6]. Addition of hydrophobic dextran derivatives leads to strings of vesicles which through the introduction of hydrophobic chains only to the outer surface re-establish the curvature in two dimensions, removed previously in converting spherical vesicles to cylinders.

Fluid lipid tubules are also found in natural systems, i.e. on the surface of leaves, on pine needles or barley leaves and are not formed by swelling with water, but are extruded from the lipophilic wax cuticula. Their purpose is presumably to filter off dust particles, which clog the transpiration pores on leaf surfaces. In pine trees, the main component is 10-nonacosanol and it is unknown whether the natural product is a pure enantiomer or a racemate. All attempts to differentiate the two alkyl chains with chiral adducts or shift reagents failed. Proof, however, has been brought forward, that only the natural product as well as the synthetic *S*-enantiomer produced tubules upon evaporation of chloroform solutions, whereas the corresponding racemate produced platelets. This finding has been explained with an ordering of long-to-short chains within a bilayer and a repulsion between hydroxy groups which appear in the same plane[7] (Figure 5.3 on page 102). Synkinesis of a curved bilayer was thus the only sign of chirality of the monomer. This most simple example of the "chiral bilayer effect" (see Figure 5.15).

[4] I. Sakurai, T. Karvabura, A. Dakurai, A. Kegami, T. Setoi, *Mol. Cryst. Liq. Cryst.*, **1985**, *130*, 203–222
[5] H. Ringsdorf, B. Schlarb, J. Venzmer, *Angew. Chem.*, **1988**, *100*, 117; *Angew. Chem., Int. Ed. Engl.*, **1988**, *27*, 113
[6] E. Kuchinka, Diplomarbeit, Universität Mainz, **1986**
[7] J.-H. Fuhrhop, T. Bedurke, A. Hahn, S. Grund, J. Gatzmann, M. Riederer, *Angew. Chem.*, **1994**, *106*, 351; *Angew. Chem., Int. Ed. Engl.*, **1994**, *33*, 350

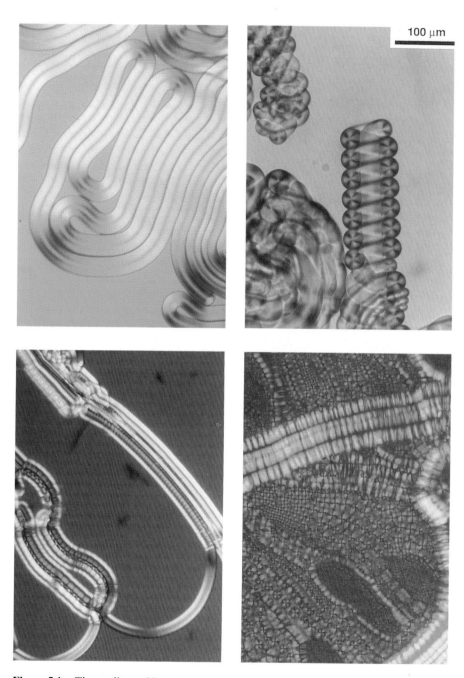

Figure 5.1 *The swelling of lecithin crystallites in water produces "myelin figures" composed of many hydrated molecular bilayers. The process depicted in the four pictures (kindly provided by Prof. I. Sakurai) takes a few hours.*

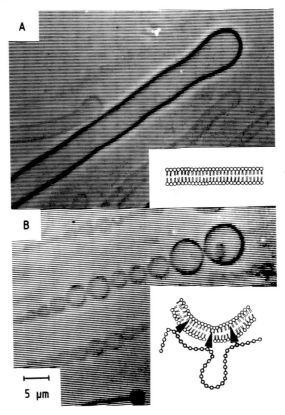

Figure 5.2 *Changes of morphology of a lecithin tubule (A) after addition of a dextran polymer with hydrophobic side-chains (B)[5]. With the kind permission of Prof. H. Ringsdorf.*

5.3 Fluid Micellar Threads

Multilayered scrolls, which are commonly formed from water-insoluble lipids upon swelling, offer no possibilities for supramolecular ordering and most of their molecular constituents have no contact with water. Furthermore, the position of dissolved or surface-attached functional molecules can neither be fixed nor determined with any certainty. Thinner surfactant rods are more promising.

A particularly long-lived type of micellar rod with very high length-to-diameter ratios can be obtained in 10^{-3}–10^{-2} M clear aqueous suspensions of cetyltrimethylammonium bromide in the presence of equimolar amounts of salicylic acid[8,9]. These rods have a diameter of approximately 12 nm (Figure

---

[8] H. Hoffmann, G. Ebert, *Angew. Chem.*, **1988**, *100*, 933; *Angew. Chem., Int. Ed. Engl.*, **1988**, *27*, 902

[9] Y. Saikaigudin, T. Shikata, H. Urakami, A. Tamura, H. Hirata, *J. Electron Microsc.*, **1987**, *36*, 168

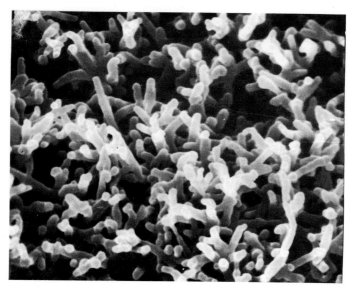

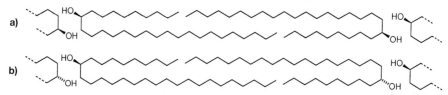

Figure 5.3 *Scanning electron microscopy image of wax tubes in their natural environment in pine needle stomata. Similar wax tubes were obtained from chloroform solutions of natural nonacosan-10-ol of unknown sterochemistry and of synthetic S-nonacosan-10-ol. The synthetic racemate only produced platelets. The curvature effect was traced back to the packing of all OH groups on one side of the bilayer (a), which is not found in the racemate (b).*

5.4a), which corresponds to three molecular bilayers. Several holes are apparent in the electron micrographs of the fibre which point to separations into bundles of single fibres. It might well be that this fibre is made of three or four micellar rods. It is known that salicylic acid, *o*- and *p*-iodophenol, and *p*-toluic acid have strong fibre-inducing effects in a 1:1 molar ratio in respect to CTAB and at concentrations between 10^{-3} and 10^{-2} M. Other hydroxybenzoic acids, diphenols, etc., have less effect or none at all. However, in the presence of 1 M sodium chloride, CTAB forms similar rods. Fibre formation from micelles therefore depends on the action of phenolic entrapments or salting effects. ^{1}H NMR spectroscopy showed that sodium salicylate is immobilized and binds to the ammonium head groups[10]. It seems likely that the polarizable phenol molecules form stacks, aligning the cetyltrimethylammonium chains. Uniform

---

[10] C. Manohar, U.R.K. Rao, B.S. Valaulikar, R.M. Iyer, *J. Chem. Soc., Chem. Commun.*, **1986**, 379

rods of 5 nm diameter have also been characterized by cryoelectron microscopy (Figure 5.4b)[11], herewith corresponding to a molecular bilayer.

More work is required in order to clarify the molecular structure of these fascinating molecular assemblies which seem to be on the borderline between fluid and solid micellar rods. Their formation develops through a certain type of precipitation, also typical for solid micellar fibres. However, the binding forces between the head group molecules (tetraalkylammonium and phenol) are weak, meaning that the fibres are not as stiff and uniform as the crystalline fibres described later. Aqueous suspensions of the described fibres dissolve massive amounts of small hydrocarbon molecules, e.g. 20 mol % of hexane, but the dissolving of hydrophobic porphyrins in them has not yet been achieved.

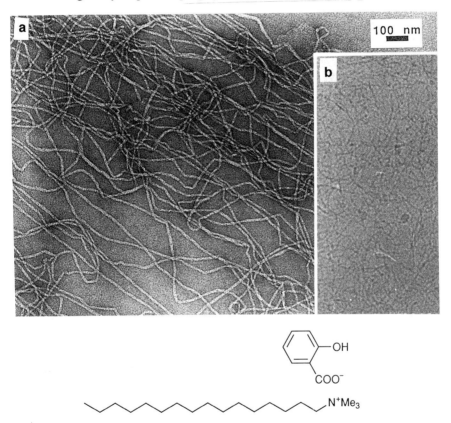

Figure 5.4 *Electron micrographs of fluid micellar threads of cetyltrimethylammonium salicylate.*
a) 12 nm threads, negatively stained with uranyl acetate, kindly provided by Prof. H. Hirata[9]. b) Insert: 5 nm threads. Preparation in a controlled environment chamber and cryomicroscopy in vitreous ice; kindly provided by Dr. C. Böttcher.

---

[11] T.M. Clausen, P.K. Vinson, J.R. Minter, H.T. Davis, J. Talmon, W.G. Miller, *J. Phys. Chem.*, **1992**, *96*, 474

It may be that the polar and disordered spherical micelles and vesicle surfaces are the only parts of molecular assemblies which provide strong solvation power. More ordered structures, even if still liquid, may not have such an effect.

The mono-tetramethylammonium salt of 2-dodecylmalonic acid also forms rod-like micelles as seen by their viscoelasticity[12]. The concentration range here is 1.4–13 mM. Light microscopy also reveals extremely long threads in sonicated sodium myristate solutions at millimolar concentrations[13].

5.4 Amide Hydrogen Bond Chains

The liquid structure of the fibres can be solidified by linear hydrogen bond chains which may run parallel or perpendicular to the fibre axis. The strongest and most prominent type of chain-like bonding interaction is provided by the amide group, which has been extensively studied in helical peptides and proteins. Here, a few characteristics of interest are summarized before turning to solid fibres. The assumption is that other hydrogen bond chains between $COOH\cdots{}^-OOC$ and $OPO_3H\cdots{}^-O_3PO-$, for example, are of a similar character.

Amide hydrogen bonds are ubiquitous between main chain atoms in proteins. The empirical length between the nitrogen atom which donates the proton and the accepting carbonyl oxygen is usually 2.9 ± 0.3 Å, with N, H and O tending to form a straight line. A deviation of 20° from 0° reduces the binding energy[14] by about 10%, resembling the magnitude of deviations observed in proteins[15]. The linearity of the hydrogen bond should not be confused with the alignment of the constituting dipoles $N-H\cdots O=C$. Misalignment of these dipoles results in much lower energy reduction of the hydrogen bond. *Ab initio* calculations also indicated that the energy surface of $NH\cdots OC$ hydrogen bonding is soft in comparison to the $NH\cdots O$ and $H\cdots OC$ angle variations[16].

Thermodynamic measurements indicated that the formation of **amide–amide hydrogen bonds in water is largely entropy driven**[17,18]. This cannot be attributed to a gain in the number of hydrogen bonds in the dimer, but instead depends on **the release of 3–4 water molecules** which are ordered ice-like in a solvated amide group. **In CCl_4, on the other hand, the driving force came directly from the negative free energy change of amide–amide hydrogen bond formation.** Altogether, amide–amide hydrogen bond formation is highly favourable in both hydrophobic and hydrophilic environments ($\Delta G = -28 \pm 4$ kJ/mol), e.g. on the surfaces of a micellar fibre or in its interior.

In protein crystal structures, ordered water molecules were frequently observed at instances where α-helices bend or fold. Molecular dynamic simu-

---

[12] J.C. Brackman, J.B.F.N. Engberts, *Langmuir*, **1991**, *7*, 46
[13] O. Träger, J.-H. Fuhrhop, to be published
[14] G.C. Pimentel, A.L. McClellan, The Hydrogen Bond, Freeman, London, **1960**
[15] P.C. Moews, R.H. Kretzinger, *J. Mol. Biol.*, **1975**, *91*, 201
[16] J.J. Novoa, M.-H. Whangbo, *J. Am. Chem. Soc.*, **1991**, *113*, 9017
[17] A.J. Doig, D.H. Williams, *J. Am. Chem. Soc.*, **1992**, *114*, 338
[18] F.M. Di Capua, S. Swaminathan, D.L. Beveridge, *J. Am. Chem. Soc.*, **1991**, *113*, 6145

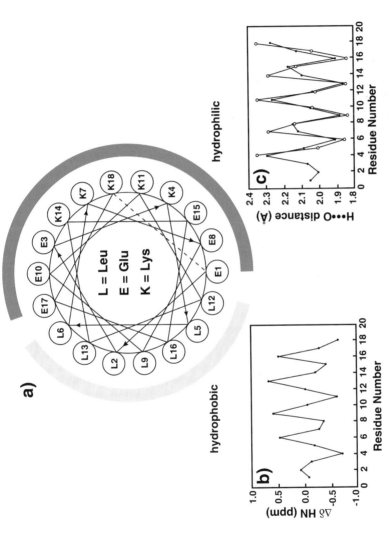

Figure 5.5 *A) Amino acid sequence of a designed amphiphilic α-helical peptide LL9 represented as a helical wheel. The hydrophobic and hydrophilic faces are observed on opposing sides. B) Plot of the difference between the chemical shifts of the amide proton of a particular residue in the peptide and of that in a random-coil conformation. C) Comparisons of the distance from the carbonyl oxygen to the amide proton obtained from the chemical shift (●) and computer modelling of a curved (!) helix (○)[20].*

lations proved that the extension of the amide hydrogen bonds by insertion of precisely ordered water molecules may cause and stabilize the bending of helices. A survey of NMR chemical shift data of amide hydrogens in peptides and proteins produced many examples showing a 3–4 repeat periodicity[19]. This was traced back[19,20] to regular changes of hydrophobic amino acids buried inside the helix and hydrophilic, solvent-exposed amino acids in amphipathic α-helical peptides. Hydrogen bonds amid buried carbonyl groups were 0.1–0.2 Å shorter than those at the outside and could be the explanation for chemical shift differences (Figure 5.5)[20]. The periodicity arises from curvature (bending) of the helices, or, if bending is not observed, one may deduce a more hydrophobic face and a more hydrophilic face on opposite sides.

5.5 Solid Molecular Mono- and Bilayer Twisted Ribbons and Tubules

Spherical micelles and vesicles are formed from amphiphiles and bolaamphiphiles by the solvophobic effect and are protected against crystallization by head group repulsion. What happens if the head groups carry secondary amide groups which have an inborn and irresistible drive to form linear hydrogen bond chains in polar and apolar environments (see section 5.4)? Chains will be formed, of course. The usual result will be vesicular tubules, as in the case of amphiphiles with a low cmc (typically $< 10^{-5}$ M) and thinner micellar rods in the case of amphiphiles with a relatively high cmc (typically 10^{-4}–10^{-2} M).

Aqueous dispersions of the amide **1** of glutamic acid with two ester alkyl chains[21] are beautiful examples on the light microscopic scale. Upon ultrasonication, they produce typical bilayer vesicles which slowly rearrange to form twisted bilayer ribbons. Electron micrographs present a static picture of these ribbons and show ill-defined twisted ribbons[22]. When the same amphiphile **1** was dispersed in water by gently shaking by hand, only flexible filaments and a few vesicles with diameters of approximately 10 μm were observed[21]. No defined supramolecular structure whatsoever appeared immediately, but after several hours, the shapeless filaments rearranged completely to form 10–100 μm long helices with a diameter of around 1 μm and a pitch of 3.2 μm. This process is comparable to 3D crystallization. One month (!) later, the helical bilayer tapes had widened to a point where closed tubules could be formed (Figures 5.6 and 5.7). Enantiomeric glutamic acid amides **1** formed mirror image helices and the racemate only gave non-helical fibres. The helical and tubular superstructures were stable only at temperatures below the phase transition of the bilayers ($T_c = 34$ °C). At 40–50 °C, the crystalline order of the bilayer was destroyed and the amide hydrogen bond chain broke, with hydration enforcing the re-formation of spherical vesicles (Figure 5.6). This process is

---

[19] I.D. Kuntz, P.A. Kosen, E.C. Craig, *J. Am. Chem. Soc.*, **1991**, *113*, 1406
[20] N.E. Zhou, B.-Y. Zhu, B.D. Sykes, R.S. Hodges, *J. Am. Chem. Soc.*, **1992**, *114*, 4320
[21] N. Nakashima, S. Asakuma, T. Kunitake, *J. Am. Chem. Soc.*, **1985**, *107*, 509
[22] N. Nakashima, S. Asakuma, T. Kunitake, H. Hotani, *Chem. Lett.*, **1984**, 227

Binding Interactions in Micellar and Vesicular Fibres

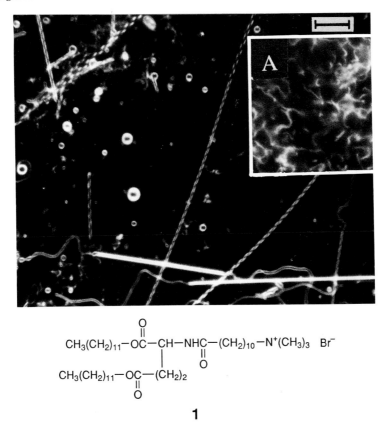

Figure 5.6 *From ill-defined filaments in aqueous suspensions (insert A) grow vesicles, helices and tubules in aqueous suspensions of the triple-chain L-glutamic acid derivatives displayed[21]. Scalebar = 10 μm. The photographs were kindly provided by Prof. T. Kunitake.*

reversible, i.e., upon cooling, helical fibres grew slowly. One can thus say that spherical micelles or vesicles are the liquid products of the melting process of solid bilayer fibres.

These fibres are solid-like and should not be confused with the fluid myelin figures and their helical precursors obtained upon the swelling of lecithin crystals (see Figure 5.1). The fluid structures flow and change their shape and width constantly, whereas the "solid types" simply widen after addition of more material. Once a crystalline fibre is formed it adds material to the highly curved edges, much less to the more planar bilayer surfaces (Figure 5.7).

An L-glutamic triamide with a pyridinium head group and two 2,4-hexadienoyl groups also produced twisted ribbons which, with ageing, closed to tubules[23]. Similar tubules were also obtained upon UV irradiation and polymerization. The CD spectrum of surface adsorbed methyl orange[23,24], however,

---

[23] H. Ihara, M. Takafuji, C. Hirayama, D.F. O'Brien, *Langmuir*, **1992**, *8*, 1548
[24] N. Nakashima, H. Fukushima, T. Kunitake, *Chem. Lett.*, **1981**, 1207

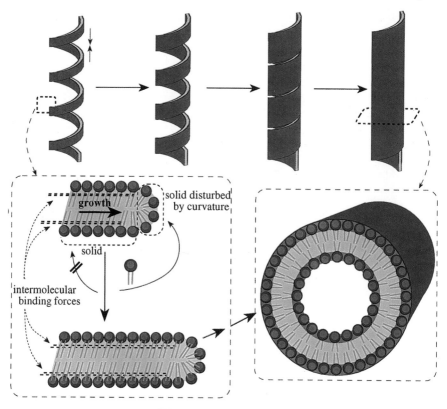

Figure 5.7 *Solid bilayer helices slowly add material to the curved edges and finally form tubules. Neither the pitch nor the width of the molecular assembly change significantly.*

disappeared upon polymerization. Bivalent counterions may have a similar effect[25], namely to destroy CD effects in water or to form water-insoluble complexes which may produce very intense CD spectra upon re-dissolution in organic solvents[26]. The helical superstructures of azobenzene containing amphiphiles cannot only be studied by CD spectra, but sometimes also reveal themselves in UV spectra. **2**, for example, forms such helical fibres in water and absorbs at 350 nm when the number of methylene groups is even and at 320 nm when uneven (Figure 5.8). The chromophore orientation is obviously more tilted (less stacking) in assemblies of even-numbered alkyl chains as in this case[27].

Single chain amides with amino acid head groups and alkyl chains of various lengths have also been shown to form helical rods, twisted ribbons and straight tubules[28]. These assemblies show no regular pitch in electron micrographs and

[25] N. Nakashima, R. Ando, H. Fukushima, T. Kunitake, *Chem. Lett.*, **1985**, 1503
[26] J.-M. Kim, T. Kunitake, *Chem. Lett.*, **1989**, 959
[27] N. Yamada, M. Kawasaki, *J. Chem. Soc., Chem. Commun.*, **1990**, 568
[28] T. Imae, Y. Takahashi, H. Muramatsu, *J. Am. Chem. Soc.*, **1992**, *114*, 3414

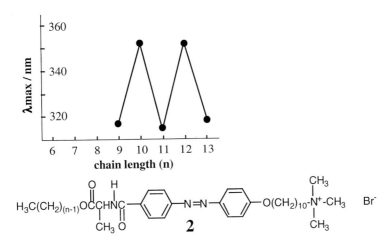

Figure 5.8 *Wavelengths of a UV absorption of chromophores in micellar fibres change periodically with even and odd numbers n of methylene groups (see text)*[27].

no light microscopic studies have been reported. Helices have, however, been characterized for N-dodecanoyl-L-alaninate in liquid crystalline material, meaning at very high concentrations (> 20 wt %). Again no microscopical supramolecular structures were observed in such concentrated preparations but they allowed laser diffraction and solid state NMR spectroscopy. The pitch was found to be 1.5 μm (!) in 30% (w/w) alaninate solutions in decanol/water 1:10 and grew to 5.3 μm in 23% (w/w) solutions[29]. Deuterium quadrupole splittings of an alaninate deuterated at NH and the CH_2 group next to the amide link, markedly decreased with an increase in concentration. It was concluded from these data that in the more concentrated solution, the chiral centre is no longer effectively shielded by the co-solvent decanol, which leads to more concerted *gauche* rotations within the liquid crystal. The smaller pitch of the more concentrated liquid crystal is therefore caused by a higher percentage of *gauche* conformations. As we shall see later (section 5.6) **crystalline** helical **bilayers with a pitch of 23 nm contain no molecules with *gauche* conformations** in the oligomethylene chains.

Raising the temperature resulted in an increasing pitch and a corresponding reduction in quadrupole splitting. This, however, simply indicates an increase in motion at the terminal positions of the amphiphile and thereby the average intermolecular distance along the helical axis. Variation in temperature (292–310 K) has, in contrast to the changes in concentration, only a minimal effect on the average conformation of the head group itself. A pitch between 1500 and 5300 nm is, however, almost of macroscopic dimensions. An explanation by molecular conformations must remain a statistical one until detailed CPMAS ^{13}C-NMR studies have been performed (see section 5.6).

[29] A.S. Tracey, X. Zhang, *J. Phys. Chem.*, **1992**, *96*, 3889

Curvature rises dramatically on application of non-symmetric bola-amphiphiles with one amino acid head group, e.g. lysine, and one amino head group. The long-chain diamide **3** dissolves up to 2×10^{-4} mol L^{-1} in water at pH 5.5 and precipitates at pH 10.5. Molecular monolayers in the form of long tubular vesicles were observed[30], their inner diameter being approximately 50 nm and their membrane thickness of 4.4 nm corresponding to the length of one molecule. The aqueous centres of the tubules were stained with phosphotungstate and were so stable that they did not collapse on evacuation (Figure 5.9). Furthermore, the water-suspended tubules could be filtered off, lyophilized and dried without being destroyed. The dried material weighed approximately the same as the applied material. The synkinetic tubule yield is therefore

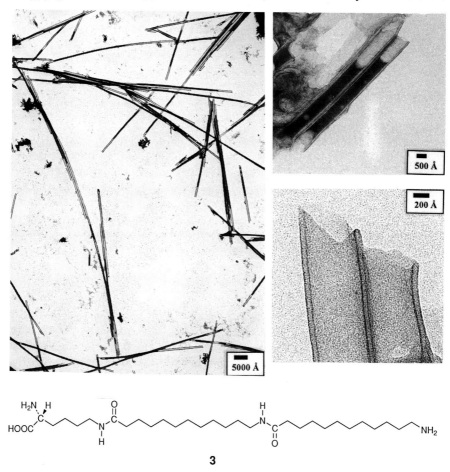

3

Figure 5.9 *Electron micrographs of vesicular tubules made of bolaamphiphile **3**, negatively stained with phosphotungstate. The stain also enters the inner part of the tubes (upper right). The thickness of the membrane is 4.5 nm (lower right), which corresponds to the molecular length of **3**.*

[30] J.-H. Fuhrhop, D. Spiroski, C. Böttcher, *J. Am. Chem. Soc.*, **1993**, *115*, 1600

almost quantitative and the intact monolayers can be stored in dry state for several weeks. Re-suspended in water, they produce exactly the same electron micrographs as if freshly prepared.

Solid ribbons and tubules, of course, can also be made with head groups other than amino acids, prominent being phospholipids with two polymerizable acetylene units, e.g. each of the two side-chains in **4**,[31] and phospholipids with a nucleotide head group, e.g. **5**.[32,33] Both types of compound are extremely water-insoluble and must be sonicated in order to form vesicles. These vesicles then grow to form multilayered tubules and/or twisted ribbons with length-diameter ratios of about 10. Multilayered tubules of light-microscopic dimensions from **4** were also obtained in acetonitrile solution[34] or were plated with nickel[35] in water. This is the first example of a combined organic/inorganic synkinesis. UV irradiation of the fibres produced polymers.

A most interesting phospholipid–nucleoside conjugate of the type **5** was an adenosine derivative which produces bundles of single helical strands with a diameter of around 5 nm and a helical pitch of about 10 nm. Image analysis[33], however, gave no uniform picture. The other phospholipid–nucleoside analogues assembled to form "cigar-type" scrolls. The clearest electron microscopic images of scrolls were obtained from *N*-octyl-D-mannonamide **6**.[36] These assemblies are transparent in negatively stained probes and can be broken smoothly. The multilayered roll of a single bilayer sheet thus becomes visible. Figure 5.10 depicts the bulkiness and rigidity of multilayered tubules very clearly. As the majority of the monomers are hidden away from the inner and outer surfaces, the membrane character of single mono- or bilayer structures is lost. Diacetylene derivatives of **6** produce the same type of scrolls and polymerize upon UV irradiation, whereby no change of the scrolls occurs[37].

---

[31] J.H. Georger, A. Singh, R.P. Price, J.M. Schnur, P. Yager, P.E. Schoen, *J. Am. Chem. Soc.*, **1987**, *109*, 6169
[32] H. Yanagawa, Y. Ogawa, H. Furuta, K. Tsuno, *J. Am. Chem. Soc.*, **1989**, *111*, 4567
[33] Y. Itojima, Y. Ogawa, K. Tsuno, N. Handa, H. Yanagawa, *Biochemistry*, **1992**, *31*, 4757
[34] A.S. Rudolph, J.M. Calvert, M.E. Ayers, J.M. Schnur, *J. Am. Chem. Soc.*, **1989**, *111*, 8516
[35] J.M. Schnur, R. Price, P. Schoen, P. Yager, J.M. Calvert, J. Georger, A. Singh, *Thin Solid Films*, **1987**, *152*, 181
[36] J.-H. Fuhrhop, P. Schnieder, E. Boekema, W. Helfrich, *J. Am. Chem. Soc.*, **1988**, *110*, 2861
[37] P.A. Frankel, D.F. O'Brien, *J. Am. Chem. Soc.*, **1991**, *113*, 7436

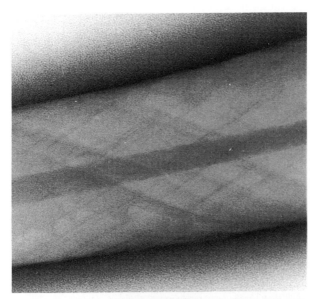

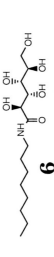

Figure 5.10 *Electron micrographs of N-octyl-D-mannonamide fibres **6**: freeze-etched (left), and negatively stained with phoshotungstate (right). Each layer has a thickness of 4 nm; the central hole has a diameter of 50 nm.*

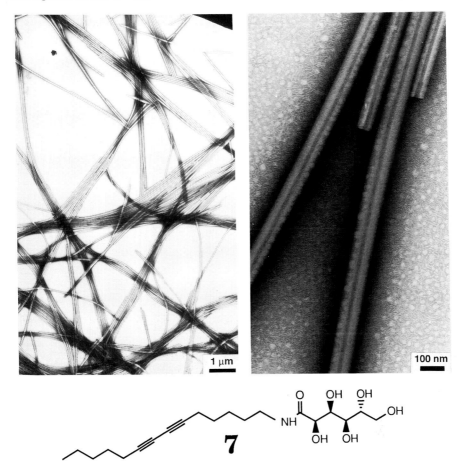

Figure 5.11 *Electron micrographs of vesicular tubules made of the diacetylenic β-gluconamide 7 (see for comparison the helical rods of N-octyl-D-gluconamide in Figure 5.13).*

Long, thin bilayer tubules were prepared from N-alkyl-D-gluconamide **7** with diacetylene units in the single chain. These compounds first formed helical ribbons on the electron microscopic scale which looked very similar to those observed by light microscopy. However, the gaps filled up so quickly that usually only the uniform bilayer membranes in tubular form could be observed (Figure 5.11)[38]. The inner diameter was normally of the order of 10 nm. In the case where only one methylene group separated the amide and diacetylene units, it was possible to polymerize the tubules to one large insoluble molecule by UV irradiation. Electron micrographs only depicted a net of polymeric threads with the diameter of a molecular bilayer.

[38] J.-H. Fuhrhop, P. Blumtritt, C. Lehmann, P. Luger, *J. Am. Chem. Soc.*, **1991**, *113*, 7437

5.6 Molecular Mono- and Bilayer Micellar Rods

Solid micellar rods consist of material-filled cylinders with the diameter of a single molecule's length as in case of bolaamphiphiles or of a molecular bilayer with single headed amphiphiles. Monolayered rods were realized with the arborol[39] **8** and the α-amino-ω-lysine bolaamphiphile[30] **9**. In both cases μm-long rods of ultimate thinness (2.0 and 2.3 nm) corresponding to the length of one molecule were obtained (Figure 5.12). Binding interactions between the polar head groups probably played a more dominant role than the hydrophobic effect on the alkyl chains. The rods were cylindrical crystals, which grew only in one direction. The bolaamphiphile rods could be isolated in dry form and re-suspended in water.

In the following sections, we give a detailed description of the rod-like micellar fibre made of N-octyl-D-gluconamide. Several principles of supramolecular synkinesis and analysis can be explained through this example; for a start, **the compound contains all structural elements** required in bioorganic chemistry: an **alkyl chain** which assembles to form lipid membranes in water, an **amide group** which introduces the ordering of monomers in linear chains similar to protein fibres and an **oligochiral head group** which induces surface chirality and possibly helicity. Furthermore, the glycon head group can be systematically varied, as all eight diastereomeric carbohydrates and three enantiomers (L-glucose, L-mannose and L-galactose) are commercially available.

N-octyl-D-gluconamide's ability to form gels and fibres of electron microscopic dimensions was discovered[40] in 1985. It was shown that up to 50 weight percent of the compound dissolved in boiling water. When a dilute solution (0.5–2%) was cooled, the compound did not precipitate, but formed opaque gels. Electron micrographs of negatively stained probes showed rope-like structures in which the single strands had a diameter of approximately 30 nm (Figure 5.13a). Nine years later, these structures have still not been isolated as a pure preparation and their construction remains unknown.

After some experimentation it was found[41] that the addition of 2% tungstate trapped **helical fibres of bimolecular thinness** (Figure 5.13b). Image analysis resulted in the picture of a bulgy double helix (Figure 5.13c,d). This model, however, did not reproduce the cross in the centre of the bulges which could be repeatedly seen in micrographs (Figure 5.13c). The preparation was continually repeated and optimized until a sample was obtained in which several hundred bulges appeared in the same orientation (Figure 5.13e). Computer averaging and image analysis then produced a detailed picture of a single motif (Figure 5.13f) and an even more detailed contour line diagram (Figure 5.13g) which

---

[39] G.R. Newkome, G.R. Barker, S. Arai, M.J. Saunders, P.S. Russo, K.J. Theriot, C.N. Moorefield, L.E. Rogers, J.E. Miller, T.R. Lieux, M.R. Murray, B. Phillips, L. Pascal, *J. Am. Chem. Soc.*, **1990**, *112*, 8458
[40] B. Pfannemüller, W. Welte, *Chem. Phys. Lipids*, **1985**, *37*, 227
[41] a) J.-H. Fuhrhop, P. Schnieder, J. Rosenberg, E. Boekema, *J. Am. Chem. Soc.*, **1987**, *109*, 3387
b) J. Köning, C. Böttcher, H. Winkler, E. Zeitler, Y. Talmon, J.-H. Fuhrhop, *J. Am. Chem. Soc.*, **1993**, *115*, 693

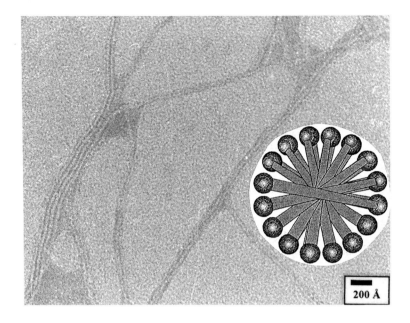

Figure 5.12 *Electron micrograph of a monolayered micellar fibre (diameter 2.5 ± 0.4 nm) made of the bolaamphiphile **9** at pH 10. Negatively stained with phoshotungstate.*

116   Chapter 5

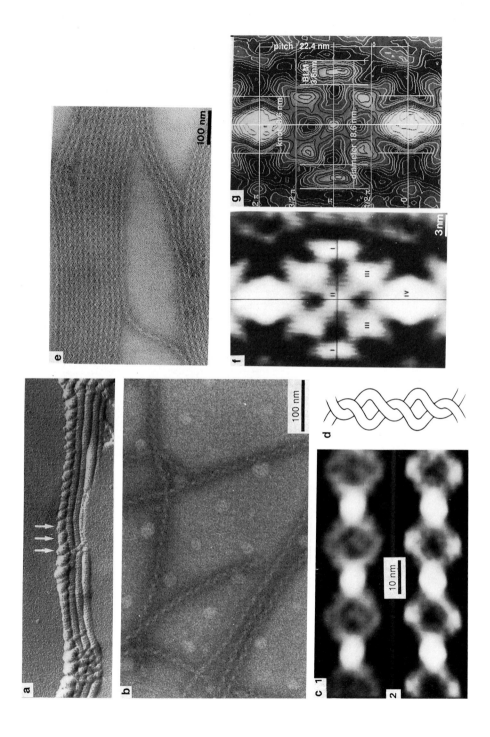

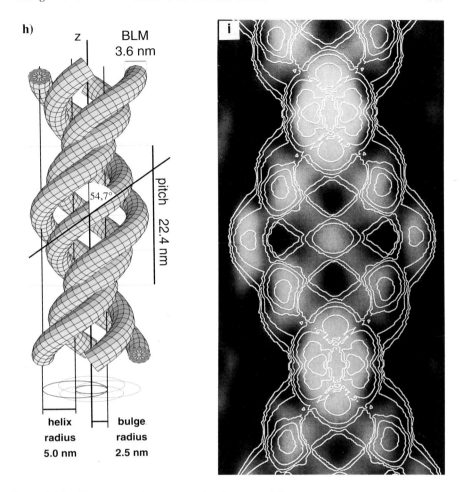

Figure 5.13 *The story of a supramolecular assembling in electron micrographs and models. For more recent results, see Frontispiece on page ix.*

a) The discovery: N-*octyl-*D-*gluconamide forms long fibres. The units (arrows) have a thickness of about 30 nm.*

b) Trapping of fibres (here: with phosphotungstate coatings at pH 4) gives uniform bulgy helices. The units have the thickness of a molecular bilayer (4 nm).

c) Image analysis of several motifs within one fibre gives better resolved pictures of (1) the octyl and (2) the dodecyl substituted amide. The knots measure four molecular lengths in each case.

d) A double helix model is proposed. It does not reproduce the shadows in the centre of the bulges in Figure c1.

e) Aligned fibres allow the image analysis of several hundred motifs.

f) As a result four helical strands are now detected and

g) a detailed contour line diagram can be derived.

h) Computer modelling on the basis of the experimental contour line diagram gives an exact 3D model. Its structure allows

i) the calculation of a contour line diagram which can be compared with the original image and the experimental contour line diagram.

could be analysed with the aid of the Mathematica® computer program. As a result, a quadruple helix was proposed (Figure 5.13h) which reproduced the cross behind the bulge perfectly and also allowed the calculation of a contour line diagram (Figure 5.13i)[42]. Apart from some flattening effects which only appeared in the experimental contour line diagram, the model and the experimental diagrams were identical. (However, very recent results suggest a quadruple helix of bilayer ribbons – see Frontispiece on page ix).

The first and foremost rule for the synkinesis of supramolecular assemblies deduced is: **Molecular assemblies can only be prepared in pure form and analysed if they are reduced to mono- or bimolecular thinness.** Multilayered assemblies will mostly be mixtures of differing assemblies containing molecules in many different environments.

The quadruple helix structure in Figure 5.13h was not predicted by synkinetic chemists. It came as a surprise and can only be understood in connection with the crystal structure of the N-octyl-D-gluconamide (see Figure 7.12). In this structure, the terminal OH-group of the primary alcohol is bent-in by a strong hydrogen bond. The surface of the crystal sheets therefore consists of hydrophobic CH_2 groups. The finding of four helices can then be explained as follows: (i) the spherical micelles in hot solution rearrange to disks upon cooling, by a cooperative formation of amide hydrogen bond chains; (ii) four such disks stick together with their hydrophobic edges and (iii) grow to rods by stacking of the hydrophobic surfaces (Figure 5.14a). In these quadruple rods, there is less hydration of the head groups at the inner surface than at the outer surface. This leads (iv) to a bending of the rods to form helices (see section 5.4). (v) The growing helical fibres develop a supramolecular dipole moment which repulse neighbouring helices. Repulsion of dipole moments is minimal if the helices have a magic angle (half tetrahedral angle) inclination against the axis. As a result, the helix winds on a helicoid minimal surface (Figure 5.14b).

This interpretation of the quadruple helix only applied to physical effects. Nevertheless, the stereoselective binding interactions between the single molecules are necessary for its formation. For example, if the enantiomeric L-gluconamide is employed, a left-handed helix is formed instead of the right-handed one. This is to be expected, but the finding that a racemic mixture of D- and L-gluconamides does not form fibres is more important. Hot solutions of both the D-enantiomer and the racemate contain spherical micelles with the same cmc, i.e. 2×10^{-3} M. Upon cooling, however, the racemate precipitates as planar platelets[40]. The question then arises as to why the pure enantiomer does not act in the same way. The answer comes from the crystal structure of the D-gluconamide which shows a head-to-tail arrangement of the crystal sheets (see Figure 7.12), whereas strong indirect evidence shows that the racemate crystallizes tail-to-tail. Subsequently, the rearrangement of the helical fibre to crystals is slow as the procedure involves an unlikely re-ordering of the micellar bilayer. Since the crystal's sheet arrangement is dominated by chirality, the stability of the fibre also critically depends on the chirality of the head group.

[42] S. Svenson, J. Köning, J.-H. Fuhrhop, *J. Phys. Chem.*, **1994**, *98*, 1022

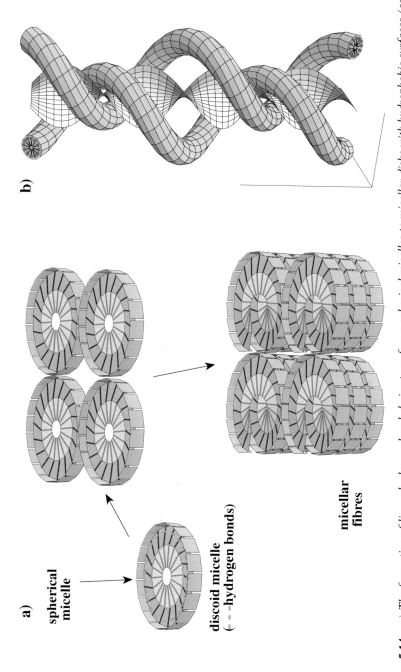

Figure 5.14 *a) The formation of linear hydrogen bond chains transforms spherical micelles to micellar disks with hydrophobic surfaces (see text). Four micellar disks assemble and grow to quadruple fibres. Negative staining with phosphotungstate plays an unknown role in this process.*
b) Dipole repulsion between the rods produces a quadruple helix which winds on a minimal surface helicoid.

In several other fibres, similar effects were observed: only the pure enantiomers produced helices of both high curvature and of a long lifetime. The racemate precipitates as platelets. This effect which dominates synkineses of many chiral (= helical) supramolecular structures has been named the **chiral bilayer effect** (Figure 5.15)[41] and could possibly play a role in biological gels made of the helical polymers and water. Take the eye ball, for example, where the gels are so long-lived because the chiral collagen and polysaccharide fibres do not crystallize (see Figure 5.29).

3D crystals and lyophilized fibres of *N*octyl-D-gluconamide and microcrystals of its racemate have therefore been analyzed by solid state CPMAS-^{13}C-NMR spectroscopy[43]. The assignment of signals was possible (i) by a comparison with NMR spectra in DMSO solution[44], in which the head group conformation was determined from coupling constants and (ii) by a comparison with the crystal structure. The orientation of crystal sheets (head-to-tail or tail-to-tail) was also obtained from differential scanning calorimetry and powder X-ray diffraction data[42,43].

It is not appropriate to outline the total structural analysis here, but we

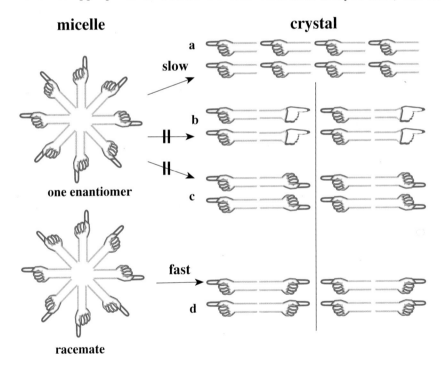

Figure 5.15 *The chiral bilayer effect. Micellar fibres are long-lived, if crystallization is connected with a slow rearrangement from tail-to-tail to head-to-tail oriented sheets. Such rearrangements occur in fibres of pure enantiomers, but not in racemates.*

---

[43] S. Svenson, Ph.D. Thesis, Freie Universität Berlin, **1993**
[44] S. Svenson, A. Schäfer, J.-H. Fuhrhop, *J. Chem. Soc., Perkin Trans. 2*, **1994**, 1023

discuss a few essential details. Heating of the head-to-tail oriented crystals of N-octyl-D-gluconamide to 100°C caused an irreversible rearrangement to a tail-to-tail arrangement. The CPMAS-^{13}C-NMR spectra of these crystals showed a shift of the terminal methyl carbon C-8' from $\delta = 15.9$ ppm to 13.9 ppm, since the head atom moved from a polar to an unpolar environment (Figure 5.16A). The crystallites of the racemic gluconamide (not shown) and the lyophilized fibres gave a signal at 13.7 ppm (Figure 5.16A) indicating a bilayer structure. Furthermore, the order of the signals for C-2 and C-4 has changed from the monolayer crystal to both the bilayer crystal and the fibre, indicating a *gauche* bend of the glucon group at C-4 in the bilayer crystal and at C-2 in the fibre. Detailed analysis yielded a $_2G^+$ bend between C-2 and C-3 (Figure 16C) in the fibre. Comparisons of solid state infrared spectra of polar and apolar crystals and apolar fibres (Figure 16B) clearly indicated the conformation of the alkyl chain: it is *all-trans* configured in all three cases. Packing constants of these chains enforce a central hole in the fibre with a diameter of about 6 Å (Figure 16D).

A conclusion can now be drawn: there is a strong 2–4 diaxial repulsion between the coaxial hydroxyl groups in the gluconamide which enforces a ***gauche*-bend at C-2, relieving strain and thus allowing hydration of the head group in micellar fibres.** The bend at C-4 causes less curvature of the bilayer and allows the formation of crystallites, but no stable 3D crystals are formed. Only the *all-trans* conformation allows optimal packing in crystal sheets with strong hydrogen bonds. Such optimal packing is, however, accompanied by the observed head-to-tail sheet rearrangement (see Figures 5.15 and 7.12).

The above methodology using crystal structures as a basis and CPMAS-^{13}C-NMR and infrared spectroscopies as major tools is generally useful for the determination of molecular conformations in molecular assemblies. It is not necessary to use a crystal structure of an amphiphile, which is often difficult to obtain. **It is sufficient to start with the crystal structure of the head group component of interest, e.g. of ethylgluconamide or gluconic acid itself.** Such simple structures can usually be taken from the literature and the CPMAS-^{13}C-NMR spectrum of the same crystals can then be measured and taken as a reference.

The CO⋯HN hydrogen bond chain is not the only grouping which introduces linear ordering in supramolecular assemblies. COO$^-$⋯HOOC and $^+$NR$_3$H⋯NR$_3$ hydrogen bond chains, which are formed at pH values close to the pK$_a$ of acids or amines, are just as effective and $^+$NH$_3$⋯$^-$OOC interactions between amino acids play a similar role (see page 110 and 124). Partially protonated or metallated phosphates can also be used. Interactions between such charged head groups cannot only be employed for producing fibres, but may also interconnect them. A striking example is the tartaric acid monoamide, which dissolves in water at pH 9 and forms huge "cloth-like" structures at the half-neutralization pH (Figure 5.17)[46].

---

[45] J.-H. Fuhrhop, S. Svenson, C. Böttcher, E. Rössler, H.-M. Vieth, *J. Am. Chem. Soc.*, **1990**, *112*, 4307
[46] J.-H. Fuhrhop, C. Demoulin, J. Rosenberg, C. Böttcher, *J. Am. Chem. Soc.*, **1990**, *112*, 2827

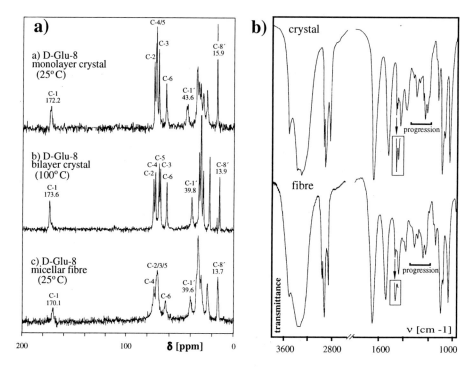

Figure 5.16 A) Solid state CPMAS-^{13}C-NMR spectra of crystalline and microcrystalline N-octyl-D-gluconamide (= D-Glu-8) materials are well-resolved. The signals have been assigned by considering *gauche effects* and by comparisons with solution state NMR spectra[43,44]. The chemical shifts in the first spectrum correspond to an all-anti conformation, shifts in the second spectrum to a $_3G^+$ *gauche bend* (b). The methyl carbon C-8' signal indicates a polar environment oriented for the monolayer crystal a), and an unpolar environment for the bilayer crystal b) and micellar fibre c). The ^{13}C signals for the lyophilized fibre in the third spectrum are somewhat broadened, but the shift of the C-2 signals can be clearly detected. This points to a G^+ conformation (b).
B) A comparison of infrared spectra of 3D crystals and lyophilized fibres reveal *identical* all-trans conformations of the CH_2 chains in both.

The fibres described above are essentially curved, one-dimensional crystals, where the growth and stability along the long axis is determined by strong linear hydrogen bond chains. Why, though, do these crystals with very high surface energies not grow quickly in three dimensions? One reason is the chiral bilayer effect already mentioned; another is that the amphiphiles dissociate from the fibre in the form of micelles and these micelles dissolve small crystallites. This effect can be strengthened by the addition of foreign surfactants, e.g. SDS, which does not integrate into the fibres, or by raising the temperature. Ultrathin solid assemblies are notoriously unstable in the refrigerator! The ultrathin *N*-octyl-D-gluconamide fibres are, for example, stable for over year

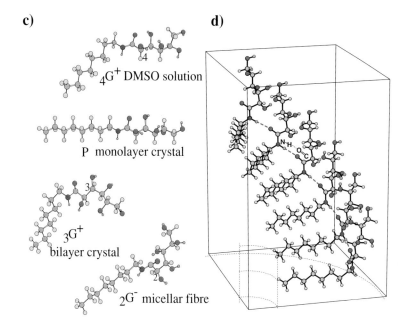

Figure 5.16 *C) Partial molecular model of the gluconamide fibre with the $_2G^+$ bend in the head group.*
D) Partial molecular model of the N-octyl-D-gluconamide fibre, showing a 0.7 nm hole in the centre (at the left of the model).

when SDS is added to the warm solution but precipitate quickly at 5°C. Presumably the SDS–gluconamide mixed micelles dissolve the crystal seeds[45]. A third reason for longevity is that bolaamphiphiles with two different head groups have a very low tendency to form 3D crystals.

5.7 Co-Crystallized Micellar Fibres

Micelles dissolve single organic molecules in aqueous media; vesicles even dissolve assemblies of organic molecules (domains), but crystalline fibres dissolve nothing at all. Guest molecules can only be introduced by co-crystallization. A synkinetic study on the co-crystallization of *N*-alkyl-glyconamides employed equimolar mixtures of D- or L-configured glucon-, mannon- and galactonamides bearing *N*-octyl or *N*-dodecyl substituents[47]. The most significant results can be summarized as follows:

(i) Amphiphiles with differing chain lengths, e.g. *N*-octyl and *N*-dodecyl gluconamides (= Glu 8, Glu 12) do not mix. Upon cooling of a hot micellar solution, the D-Glu 8 + D-Glu 12 mixture yielded bulgy clusters of micelles (Figure 5.18a) from which D-Glu 8 separated and D-Glu 12 fibres grew (not shown).

[47] J.-H. Fuhrhop, C. Böttcher, *J. Am. Chem. Soc.*, **1990**, *112*, 1768

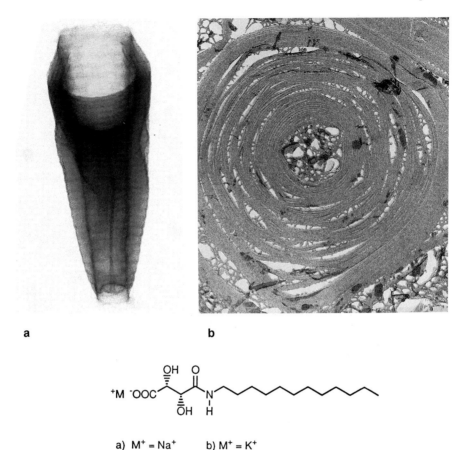

Figure 5.17 a) *Cloth-like assemblies of thousands of micellar fibres made of the N-dodecyl-D-tartaric amide shown.*
b) *Cross-section through such an assembly. Each thread has a diameter of 4 nm.*

(ii) The pseudo-racemic mixture D-Glu 8 + L-Glu 12 separated spontaneously into left- and right-handed fibres (spontaneous resolution of a pseudo-racemate; Figure 5.18b) which later recombined to form planar platelets (chiral bilayer effect; Figure 5.18c).

(iii) Fibres of high curvature (D-Glu 12) and low curvature (*N*-dodecyl-D-mannonamide = D-Man 12) either form tubular alloys (Figure 5.18d) or separate quantitatively (D-Glu 8 and L-Man 8; not shown).

(iv) Pure D-Man 18, which usually forms only platelets or scrolls (see Figure 5.10), rapidly forms right-handed (P = plus) and left-handed (M = minus) helices, if its micellar solution in SDS is cooled down (Figure 5.18e).

Co-crystalline bilayers can only be expected in tubular vesicles (Figure 5.19). Here, the curvature is not so high and linear conformations of the constituents

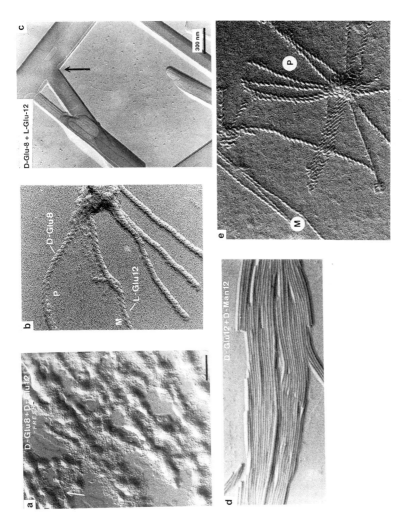

Figure 5.18 *a–d) Electron micrographs of mixtures of fibrous assemblies; see text (p. 123–4). e) The right-handed helices (P: \\\\\\) and left-handed helices (M: //////) of N-octadecyl-D-mannonamide must be caused by a bend in the mannonamide head group. Such a bend has been detected in its crystallites, and may be flexible. Therefore M (= minus) and P (= plus) helices relate to G^+ and G^- conformations.*

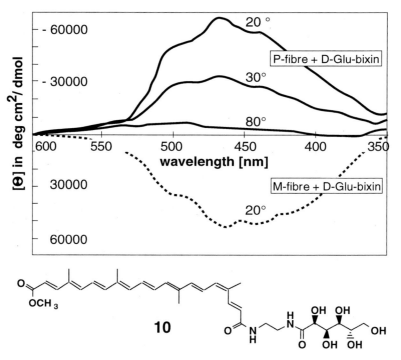

Figure 5.19 *The CD effect of **10** co-crystallized with D-gluconamide fibres (see Figure 5.13) only depends on the chirality of the fibre and not on the chirality of the chromophore's head group.*

are most probable. Fitting of hydrogen bond patterns is relatively easy. If the head groups are bent (as in the micellar gluconamide helix) the fitting of guests becomes difficult. Nevertheless it was possible to exactly integrate 0.1% (wt/wt) of a carotenoid bixin derivative into this quadruple helix. In spite of its bulky methyl substituents, bixin gluconamide **10** showed strong and inverse CD effects in both D- and L-gluconamide helices. The chirality of the bixin head group had no effect. The slender polyene thus fits perfectly into the crystalline hydrocarbon chains (see Figure 5.19), if the concentration is not too high[48].

5.8 Porphyrin and Metalloporphyrin Fibres

Attempts to co-crystallize amphiphilic porphyrins with micellar fibres failed completely. The van der Waals interactions between porphyrin planes ("stacking") are so strong that unsubstituted porphin and β-octamethylporphyrin are insoluble in practically all solvents. Porphyrins with larger substituents such as ethyl or acetic acid side chains allow face-to-face dimerization only, an action which occurs in all solvents and also probably in micelles down to a concentra-

---

[48] J.-H. Fuhrhop, M. Krull, A. Schulz, D. Möbius, *Langmuir*, **1990**, *6*, 497

tion of around 10^{-6} M (Figure 5.20b,c)[49]. The porphyrin dimers only consist of parallel oriented molecules. There are no perpendicular orientations as in the benzene dimer, where close-contact, T-shaped dimers possess the lowest free energy[50]. Porphyrin dimers produce a Soret band with a half-width of about 30 nm and practically no shift of the monomer's 400 nm Soret band. Only in the artificial *meso*-tetraphenylporphyrin, where four phenyl substituents are oriented perpendicularly to the porphyrin plane, monomolecular dissolution is enforced by sterical repulsion. *Meso*-tetraphenylporphyrins show a 15 nm wide Soret band.

The porphyrin macrocycle (as a unit of assemblies) can be taken as a rigid hydrophobic box with special packing properties, the geometric dimensions of the box being $0.7 \times 0.7 \times 0.34$ nm^3 (Figure 5.20a). As the box has a high density of π-electrons in the large periphery and a positive hole in the central cavity, it dimerizes with a lateral shift, i.e. an electron-rich pyrrole ring of one porphyrin migrates towards the centre of a partner porphyrin and stays there. Extremely stable dimers are thus formed in solution as well as in crystal structures and probably also form the unit of higher aggregates in water[51].

The ubiquitous porphyrin homodimers are only of interest in their oxidized form. The zinc octaethylporphyrin cation radical, for example, forms a diamagnetic dimer with an extremely strong charge transfer band in the near infrared (Figure 5.21). In crystalline material, all four pyrroles lie exactly on top of each other with an interplanar distance of 0.42 nm[53]. The π, π'-dimer behaves like a

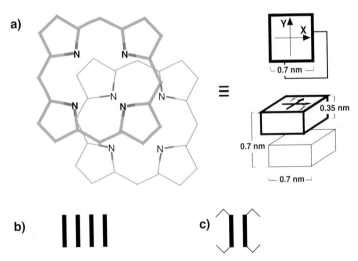

Figure 5.20 *a)* The porphyrin π-system produces a thickness of 0.35 nm of the "porphyrin box", of which the length and width measures only twice as much. Electron-rich pyrrole rings tends to combine with the electron-poor porphyrin cavity. *b)* Porphyrin forms polymer stacks, whereas *c)* octaethylporphyrin only dimerizes.

[49] J.-H. Fuhrhop, *Angew. Chem.*, **1974**, *86*, 363; *Angew. Chem., Int. Ed. Engl.*, **1974**, *13*, 321
[50] P. Linse, *J. Am. Chem. Soc.*, **1992**, *114*, 4366
[51] C.A. Hunter, J.K.M. Sanders, *J. Am. Chem. Soc.*, **1990**, *112*, 5525

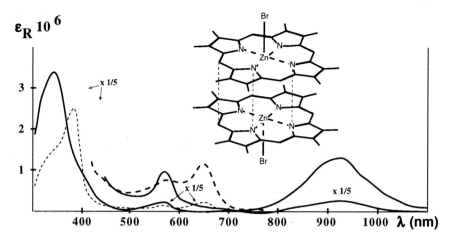

Figure 5.21 *The oxidized porphyrin radicals form very stable dimers, in which the unpaired electrons reversibly form a π–π bond.*

monomer in which the unpaired electrons form a bond with 17.5 kcal/mol binding energy[52].

The homodimers are standard; heterodimers must be synkinetized intelligently. In an earlier example, positively and negatively charged tetraphenylporphyrin (TPP) derivatives were used producing a polymeric material. Mixing aqueous or ethanolic solutions of tetrasodium *meso*-tetrakis(*p*-sulfonatophenyl)porphyrin (TPPS) and the corresponding *N,N,N*-trimethylaniliniumporphyrin (TTAP) *or* their metal complexes produced precipitates which, by elementary analysis, were shown to be 1:1 porphyrin cation–anion aggregates[54]. These aggregates are slightly soluble in water/acetone (1:1) and have a broadened, blue shifted Soret band (λ_{max} 400 nm instead of 420 nm). The dimers fluoresced strongly, but the fluorescence of free bases and of zinc complexes was quenched when paired with a copper porphyrin[55]. Formation of ZnTTAP/ZnTPPS dimers did *not* affect photoexcited triplets, meaning that the excitation energy must be located on the monomers[56]. This again indicates that TPP-type porphyrins hardly interact electronically.

Quantitative heterodimerization without precipitation was achieved between 5,10,15,20-tetrakis(*N*-methylpyridinium-4-yl)porphyrins and 4,4′,4′,4′′′-tetrasulfonated phthalocyanines in DMSO. Models show that there is practically no hindrance in the face-to-face assembly (Figure 5.22). Dramatic spectroscopic changes by exciton interactions were therefore observed upon dimer formation. In water, the same heterodimerizations developed, but self-aggregation processes competed. It was also discovered that the binding

[52] J.-H. Fuhrhop, P. Wasser, D. Riesner, D. Mauzerall, *J. Am. Chem. Soc.*, **1972**, *94*, 7996
[53] H. Song, C.A. Reed, W.R. Scheidt, *J. Am. Chem. Soc.*, **1989**, *111*, 6865
[54] E. Ojadi, R. Selzer, H. Linschitz, *J. Am. Chem. Soc.*, **1985**, *107*, 7783
[55] H.V. Willingen, U. Das, E. Ojadi, H. Linschitz, *J. Am. Chem. Soc.*, **1985**, *107*, 7784
[56] H. Segawa, H. Nishino, T. Kamikawa, K. Honda, T. Shimidzu, *Chem. Lett.*, **1989**, 1917

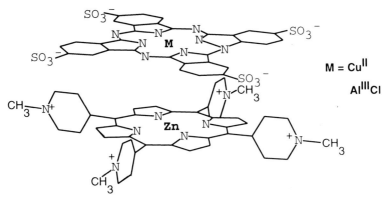

Figure 5.22 *Schematic representation of the co-facial heterodimer of a metallo-tetra-sulfonatophthalocyaninate ($M = Cu^{2+}, Al^{3+}$) and zinc meso-tetrapyridinium porphyrinate.*

strength of two metal complexes was weak in comparison to a free base-metal complex pair. Furthermore, heterotrimers of the sandwich type with two porphyrins separated by a phthalocyanine were occasionally obtained or vice versa[57,58].

Large metal ions such as thorium or uranium were also employed in order to combine a porphyrin, octaethylporphyrin (OEP), with a phthalocyanine (Pc) chromophore in face-to-face or "double-decker" orientation. U^{IV} and Th^{IV} compounds could be electrogenerated in up to seven different oxidation states. The first two reductions occurred mainly on the Pc ring while the first two oxidations of the same "heteroleptic" compounds involved primarily orbitals of the porphyrin macrocycle. In contrast, the homoleptic $(MP)_2$ and $(MPc)_2$ derivatives undergo initial one-electron oxidations and reductions which are localized not at a single ring, but involve both macrocycles[59].

In complete contrast to metalloporphyrins, the analogous metallochlorins showed no π–π attractive interactions[60] whatsoever. An important assembly occurred in the case where the keto oxygen of the isocyclic ring in chlorophylls bound to the central magnesium ion of another molecule. Dimers with an interplanar separation of 4.2–4.7 Å were formed[59]. A series of earlier papers handled the characterization of such dimers by ^{1}H-NMR and IR spectroscopies[60,61], and their behaviour has been compared with covalently connected dimers[62,63].

57 J.F. Lipskier, T.H. TranThi, *Inorg. Chem.*, **1993**, *32*, 722
58 T.H. Tran-Thi, J.F. Lipskier, D. Houde, C. Pépin, E. Keszel, J.P. Jay-Gérin, *J. Chem. Soc., Faraday Trans.*, **1992**, *88*, 2129
59 K.M. Kadish, G. Moninot, Y. Hu, D. Dubois, A. Ibnefassi, J.-M. Barbé, R. Guilard, *J. Am. Chem. Soc.*, **1993**, *115*, 8153
60 K.J. Abraham, A.E. Rowan, N.W. Smith, K.M. Smith, *J. Chem. Soc., Perkin Trans. 2*, **1993**, 1047
61 J.J. Katz and H. Scheer, in K.M. Smith (ed.), Porphyrins and Metalloporphyrins, Elsevier, Amsterdam, **1975**
62 J.E. Hunt, J.J. Katz, A. Svirmickas, J.C. Hindman, *J. Am. Chem. Soc.*, **1984**, *106*, 2242 and references therein
63 J.F. Hinton, R.D. Harpool, *J. Am. Chem. Soc.*, **1977**, *99*, 349

Bacteriochlorophyll (BChl) and bacteriopheophytin (BPhe) oligomers showed a maximal optical absorption at 860 nm in 3:1 formamide/water solution and were in a temperature dependent equilibrium with the monomers[64,65]. BChl-860 is a large oligomer which can be centrifuged. Under the electron microscope, both the magnesium complex BChl and the metal-free BPhe form short, non-characteristic cylinders with constant diameters of 15 and 40 nm and an average length of approximately 200 nm. These cylinders produce strong circular dichroism (CD) bands. It was estimated that one fibre contained about 300 BChl or 700 BPhe dimers and that the separation between the dimers was about 2 nm. Since neither the BChl nor the BPhe molecule bears strong hydrogen donating groups, it is presumably necessary to hold the assembly together with formamide molecules[64,65]. The keto group presumably acts as a proton acceptor; the phytol side-chains play the role of the solvophobic part. The whole assembly has the character of a very loose, inverse micellar strand with a helical twist.

A similar metal–substituent interaction was also realized in TPP assemblies. The zinc complex of 5-pyridyl-10,15,20-triphenylporphyrin formed a polymer in chloroform at millimolar concentration. Its equilibrium constant was calculated with a statistical relationship assumed to be 3.1×10^4. ^{1}H-NMR spectro-

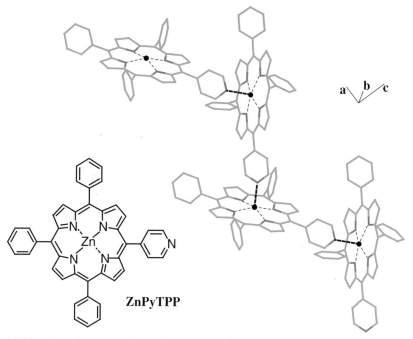

Figure 5.23 *Fibre formation by a "bis-functional" metalloporphyrin. One functional group is the central zinc ion, the other the peripheral pyridine ring.*

---

[64] J.R.E. Fisher, V. Rosenbach-Belkin, A. Scherz, *Biophys. J.*, **1990**, *58*, 461
[65] A. Scherz, V. Rosenbach-Belkin, *Proc. Natl. Acad. Sci. USA*, **1989**, *86*, 1505

scopy suggested a zig-zag chain connected by zinc–pyridine bonds (Figure 5.23). This was confirmed in detail by an X-ray crystal structure[66].

Short porphyrin and phthalocyanine "μ-oxy fibres" were obtained via dehydration of four-valent metal complexes (Si, Ge, Sn) with two hydroxy ligands[67].

The most stable and diversified porphyrin assemblies were received from protoporphyrin IX derivatives. It is characteristic that **no crystal structure of an amphiphilic porphyrin is known**. Crystal structures of esters have frequently been solved (see section 7.5), but protoporphyrin IX or similar porphyrins with a hydrophilic and a hydrophobic edge withstood all attempts at crystallization. This may be because they prefer to occur in curved fibrous assemblies.

Protoporphyrin IX dissolved in water at pH 9 and formed molecular assemblies at pH 4.5 near the pK_a of the propionic acid groups. Gel chromatography showed that the colloidal particles have molecular weights above 10^6. A particular characteristic is the splitting of the Soret band with broad peaks at 360 and 450 nm[68]. Electron micrographs, however, showed no characteristic supramolecular structures of the unstable colloidal particles[69].

Other amphiphilic porphyrins formed perfect micellar fibres in water which were frequently characterized by electron microscopy. The D-glucosamide **11**, for example, produced typical short grains with a diameter of 6–10 nm and an average length of about 200 nm. They process the invaluable advantage of remaining suspended in water for longer than a year[69]. The open-chain carbohydrate head group in **12** induces the formation of extremely long fibres which remain in suspension for weeks and even months[70]. Chiral fibres are regarded as being the most stable, a fact also associated with four-coordinated tin(IV) porphyrinates **13** which cannot stack but nevertheless assemble to very long, 6 nm wide fibres at pH 0 (Figure 5.24)[70]. Protonated chloride counterions (HCl^+—Sn) act as facial head groups above and below the porphyrin plane[70,71]. These fibres have a high tendency to precipitate[72]. The achiral porphyrin amine **14** is not only capable of forming micellar fibres, but also vesicular tubules. Protoporphyrin phosphates **15** also produce long, thin fibres.

The electronic spectra of the porphyrin solutions containing the micellar or vesicular fibres usually only show a blue-shifted Soret band between 350 and 380 nm pointing to columnar stacks. However, in addition to the short-wavelength band a long-wavelength band around 450 nm can frequently be observed (Figure 5.25A). Lateral packing of the porphyrin chromophores at the hydrophobic edges was held responsible for this band. A proposed model for a fibre with a split Soret band at 450 and 350 nm is shown in Figure 5.25B.

[66] E.B. Fleischer, A.M. Schachter, *Inorg. Chem.*, **1991**, *30*, 3763
[67] M.D. Hohol, M.W. Urban, *Polymers*, **1993**, *34*, 1995
[68] I. Inamura, K. Uchida, *Bull. Chem. Soc. Jpn.*, **1991**, *64*, 2005
[69] J.-H. Fuhrhop, C. Demoulin, C. Böttcher, J. Köning, U. Siggel, *J. Am. Chem. Soc.*, **1992**, *114*, 4159
[70] J.-H. Fuhrhop, U. Bindig, U. Siggel, *J. Chem. Soc., Chem. Commun.*, **1994**, 1583
[71] J.-H. Fuhrhop, U. Bindig, U. Siggel, submitted
[72] J.-H. Fuhrhop, U. Bindig, U. Siggel, *J. Am. Chem. Soc.*, **1993**, *115*, 11036

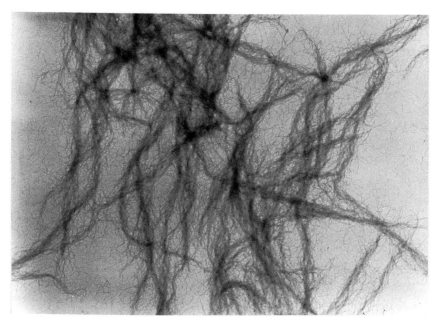

Figure 5.24 *Electron micrograph of porphyrin micellar fibres obtained from aqueous suspensions of porphyrin 13. The thin fibres have a diameter of 6 nm.*

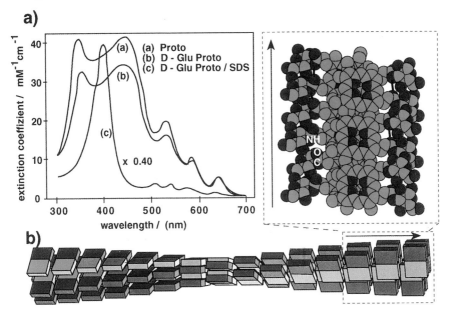

Figure 5.25 *A) Absorption spectra of aqueous solutions of (a) protoporphyrin IX (pH 7; 20% DMSO), (b) protoporphyrin bis(2-aminoglucosamide) 11 and (c) after addition of SDS to (b). B) The computer graphs show the assumed twisted assembly of tetrameric units.*

Through a change of porphyrin symmetry, substituents, central ions and axial ligands, it is therefore possible to enforce either columnar or lateral packing. The π-systems and/or central metal ions can interact either face-to-face or edge-to-edge, and in both cases, porphyrin fluorescence was completely quenched and the porphyrins became extremely light-stable. Chiral substituents caused a twisting of the rods and ribbons and large CD effects were induced. As an example, the CD spectra of the zinc and copper complexes of porphyrin ligands were reproduced (Figure 5.26). They showed opposite signs, meaning that the helical sense of the columns change from one metal to the other[71].

As expected from the extremely low fluorescence of fibres made of alkyl-substituted porphyrin amphiphiles, flash photolysis is ineffective. Nevertheless, the formation of porphyrin anion radicals was detected on a millisecond time scale and was traced back to a charge separation within the porphyrin fibre[72].

The "octopus"-porphyrin **16**, with four long alkyl chains and zwitterionic head groups on both the upper and lower surfaces of tetraphenylporphyrin, binds together in water to form fibres of monomolecular thickness[73]. Only the substituents interact, with a minimal effect on the UV/VIS spectra; most importantly the fluorescence of the fibres was strong. Such TPP-derived fibres constitute ideal "antipodes" to the bilayers whose chromophores are not widely removed from each other. Another consequence of this special environment is the reversible formation of oxygen adducts of the corresponding heme fibres.

[73] T. Komatsu, K. Nakao, H. Nishide, E. Tsuchida, *J. Chem. Soc., Chem. Commun.*, **1993**, 728

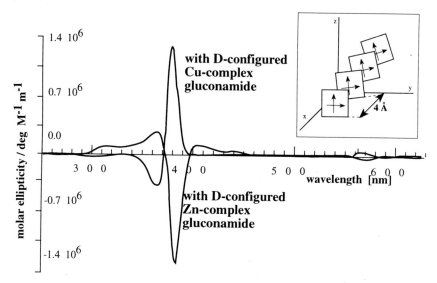

Figure 5.26 *CD spectra of fibres in water made of zinc and copper complexes of the bis(D-gluconamide) 12.*

The porphyrin fibres are relatively loosely packed polymers with either strong exciton interaction and no fluorescence (protoporphyrin derivatives) or no exciton interaction and strong fluorescence (tetraphenylporphyrin derivatives). These polymeric assemblies must be compared to the classical Jelley[74] or Scheibe[75] polymers of methine cyanine dyes in water. In these polymers a strong π–π interaction between electron donating (= d) and accepting (= a) sites[76,77] leads to very rigid trimer or tetramer units (Figure 5.27) which produce very narrow long-wavelength bands together with a strong fluorescence of the polymeric assemblies[78]. The narrowing of the linewidth of the long-wavelength band has been traced back to strong intermolecular coupling of the molecular excitations. The excitons are thereby delocalized and evenly distributed over local inhomogeneities of the polymeric assembly[79]. Obviously, the electron donor–acceptor interactions are much stronger in linear cyanine dyes than in the macrocyclic porphyrins. Electron micrographs show long rods with a minimal width of about 2.8 nm[80].

An important aspect of porphyrin assembly chemistry is the interaction of porphyrins with colloids and polymers. For years it has been recognized that chlorophyll a is practically insoluble as a monomer and nonfluorescent in hydrocarbon solvents unless a "polar activator" is present[81]. Concentrated

[74] E.E. Jelley, *Nature (London)*, **1936**, *138*, 1009; **1937**, *139*, 631
[75] G. Scheibe, *Angew. Chem.*, **1936**, *49*, 563; **1937**, *50*, 51
[76] F. Dietz, *Tetrahedron*, **1972**, *28*, 1403
[77] H. Bücher, H. Kuhn, *Chem. Phys. Lett.*, **1970**, *6*, 183
[78] A.H. Herz, *Adv. Colloid Interface Sci.*, **1977**, *8*, 237
[79] E.W. Knapp, *Chem. Phys.*, **1984**, *85*, 73
[80] E.S. Emerson, M.A. Conlin, A.E. Rosenoff, K.S. Norland, H. Rodriguez, D. Chin, G.R. Bird, *J. Phys. Chem.*, **1967**, *71*, 2396
[81] R. Livingston, W.F. Watson, J. McArdle, *J. Am. Chem. Soc.*, **1949**, *71*, 1542

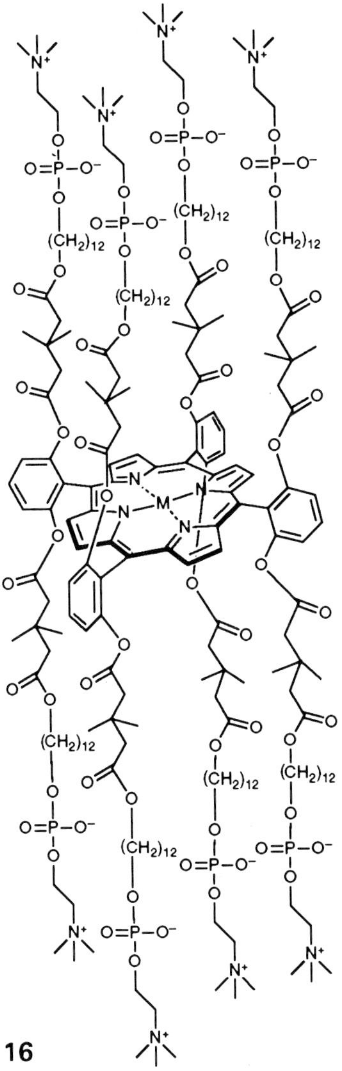

16

solutions of chlorophyll a (1.2 × 10⁻³ M) can, for example, be obtained in n-undecane containing a fifty-fold excess of N,N-dimethylmyristylamide (DMMA; 56 mM) which is heavily aggregated in this solvent. The Soret band appeared at 431 nm and fluorescence was observed[82]. If the secondary polyamide N-methylmyristylamide (MMA) was employed instead of the less polar tertiary amide, cyclic chlorophyll oligomers wrapped themselves around the polyamide. A strong absorption band for "crystalline" chlorophyll appeared at 744 nm together with a CD band at 680 nm with no observation of fluorescence[82]. Where DMMA was absent, much of the chlorophyll existed as an aggregate

---

[82] Y. Kusumoto, V. Slenthilathipan, G.R. Seely, *Photochem. Photobiol.*, **1983**, *37*, 571

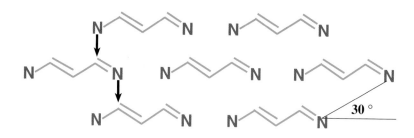

Figure 5.27 *Schematic representation of the polymeric Jelley or Scheibe assemblies of isocyanine dyes. An imine carbon atom acts as electron acceptor (a), an enamine nitrogen as donor (d). If the angle α is much below 54°, long wavelength polymer bands are observed. The rigid polymer fluoresces.*

absorbing at 741 nm. Here probably lies a solution to the problem of dissolving aromatic systems in apolar membrane systems: polar assemblies of activators with hydrophobic edges must first be introduced into the membranes. This is more or less the same solution which is practised in nature, where chromophores are squeezed in between helical membrane proteins.

Sometimes smaller porphyrin assemblies could be reversibly dissolved via molecular complex formation. The protoporphyrin-bis-amide **17** with two *m*-phenylboronic substituents dissolves in 1:30 DMSO/water mixtures, but is heavily aggregated in this medium. The Soret band's intensity was only the half of that in pure DMSO solution and the fluorescence was almost nil. However, upon addition of 10^{-2} M fructose the carbohydrate was bound as a molecular complex and the porphyrin became more water-soluble whereby the fluorecence increased drastically[83]. Other monosaccharides had lesser effects.

Protoporphyrin carries two vinyl groups and can therefore be copolymerized with other vinyl monomers, e.g. styrene. The product usually contains isolated porphyrin units in statistical arrangements and has a totally hydrophobic environment. Several properties of such statistical assemblies differ from those in homogeneous solution. Iron(II) porphyrins in polystyrene copolymers which contain some imidazole, reversibly add oxygen without being oxidized to iron(III) porphyrin[84]. Although these porphyrins are connected to a polymeric backbone they behave as isolated monomers. This situation could also be found in haemoglobins. Analogous, water-soluble porphyrin copolymers, e.g. with poly(1-vinyl-2-pyrrolidone) formed an oxygen adduct even in aqueous media[85]. Solution of hindered hemes in vesicles and vesicular hemes also behaved as

---

[83] H. Murakami, T. Nagasaki, I. Hamachi, S. Shinkai, *Tetrahedron Lett.*, **1993**, 6273

[84] J.-H. Fuhrhop, S. Besecke, W. Voigt, J. Ernst, J. Subramanian, *Makromol. Chem.*, **1977**, *178*, 1621

17

liquid-like porphyrin assemblies and reversibly added oxygen molecules[86,87]. Extremely useful blood substitutes were thus synkinetized.

The adsorption of *meso*-tetrapyridylporphyrins (and similar cationic porphyrins of the tetraphenylporphyrin kind) to both natural DNA and synthetic homonucleotides have been studied over many years. All the porphyrins at AT sites were bound externally, involving some degree of overlap between the porphyrin and the 4 bases of the duplex[88]. The binding of zinc tetramethylpyridinium porphyrinate (ZnTMPyP) to poly[(dA–dT)$_2$] is cooperative, suggesting that the porphyrin induces an allosteric transition in the DNA which may widen the minor groove and create an optimum binding site[89]. Modelling showed that unperturbed minor grooves in AT-rich regions appeared to be too small for binding porphyrins; end-on binding in the major groove would not explain the observed cooperative allosteric transitions[89,90]. Intercalation occurred only at the GC sites (Figure 5.28)[88]. Equilibrium constants were in the order of 10^{-4} M^{-1}; copper and nickel complexes showed negative CD bands with molar ellipticities around 5×10^4. Axially ligated metalloporphyrins do not intercalate into DNA helices. This is equally so for vanadyl *meso*-tetramethylpyridinium porphyrinate, which binds strongly to [poly(dA–dT)$_2$]$_2$ surfaces. A CD band with a molar ellipticity of -5.5×10^5 deg cm^2 dmol^{-1} at 441 nm was observed. The vanadyl porphyrin is therefore a useful paramagnetic probe for AT regions of DNA[91].

[85] E. Tsuchida, H. Nishide, H. Yokoyama, R. Young, C.K. Chang, *Chem. Lett.*, **1984**, 997
[86] E. Tsuchida, H. Nishide, M. Yuasa, T. Babe, M. Fukuzumi, *Macromolecules*, **1989**, *22*, 66
[87] E. Tsuchida, T. Komatsu, K. Arai, H. Nishide, *J. Chem. Soc., Chem. Commun.*, **1993**, 730
[88] R.F. Pasternack, E.J. Gibbs, J.J. Villafranca, *Biochemistry*, **1983**, *22*, 5409
[89] R.F. Pasternack, E.J. Gibbs, J.J. Villafranca, *Biochemistry*, **1983**, *22*, 2406
[90] L.G. Marzilli, *New J. Chem.*, **1990**, *14*, 409
[91] M. Lin, M. Lee, K.R. Yue, L.G. Marzilli, *Inorg. Chem.*, **1993**, *32*, 3217

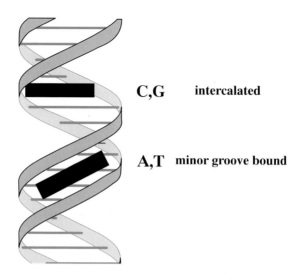

Figure 5.28 *Cationic porphyrins are preferably bound in the minor cleft (AT-rich sequences) and are intercalated in CG-rich sequences.*

The nucleic acid bases do *not* ligate to axial sites of the metalloporphyrins. Nevertheless, stability constants were larger for purine bases than pyrimidine bases, with a given porphyrin derivative[92]. The ethidium ion E^+ showed an increase of fluorescence by a factor of about 5 when intercalated into DNA. This effect was traced back (i) to an isolation of E^+ from the solvent and (ii) to the accompanying elimination of proton transfers to the excited state of E^+. At a distance of 25–30 Å cationic porphyrins P^+ are capable of virtually totally quenching the ethidium fluorescence[93]. An extremely efficient transition state dipole coupling within the E^+–DNA–P^+ complex has been held responsible for this extraordinary long range energy transfer of the Förster type.

The finding that dicationic 5,15-diphenyl-10,20-dipyridinium (= *trans*-DPyP) porphyrins show a much greater tendency to aggregate with each other than the tetracationic analogues is most interesting. Upon binding to nucleic acids, these porphyrins form long-range structures giving more intense CD signals whose profiles report the helical sense of the DNA, namely Y(+) and Y(−) DNA[94–97] (molar ellipticities: $2–3 \times 10^5$). Porphyrin–porphyrin interactions have been held responsible for these intense CD bands. Occasional medical applications

---

[92] J.A. Strickland, L.B. Marzilli, K.M. Gay, W.D. Wilson, *Biochemistry*, **1988**, *27*, 8870
[93] R.F. Pasternack, M. Caccam, B. Keogh, T.A. Stephenson, A.P. Williams, E.J. Gibbs, *J. Am. Chem. Soc.*, **1991**, *113*, 6835
[94] R.F. Pasternack, E.J. Gibbs, J.J. Villafranca, *Biochemistry*, **1983**, *22*, 2406
[95] R.F. Pasternack, E.J. Gibbs, A. Gaudemer, A. Antebi, S. Bassner, L. De Poy, D.H. Turner, A. Williams, F. Laplace, M.H. Lansard, C. Merienne, M. Perrée-Fauvet, *J. Am. Chem. Soc.*, **1985**, *107*, 8179
[96] R.F. Pasternack, C. Bustamente, P.J. Collings, A. Giannetto, E.J. Gibbs, *J. Am. Chem. Soc.*, **1993**, *115*, 5393
[97] R.F. Pasternack, A. Giannetto, P. Pagano, E.J. Gibbs, *J. Am. Chem. Soc.*, **1991**, *113*, 7799

and biological effects of DNA–porphyrin interactions result from the redox and photochemical cleavage reactions of DNA with adsorbed porphyrins[90,91]. If the tetraphenylporphyrin derivative carries an ellipticine substituent **18** which easily intercalates into DNA, then highly fluorescent complexes are formed, indicating fully separated porphyrin units[98]. Such porphyrin–nucleic acid complexes are of medical interest, as certain porphyrins can be used as anti-tumour active photosensitizers and probably possess the capability to break DNA strands upon irradiation[92,97].

18

R = -O(CH$_2$)$_3$NHCO(CH$_2$)$_5$-

A tetraanionic porphyrin with four *p*-sulfonatophenyl substituents on the methine bridges (PTS) binds to cationic polyvinylpyrrolidone (PVP) with an equilibrium constant of 1.4×10^{-7} M^{-1}. As with all other *meso*-tetraphenyl derivatives, there is no sign of porphyrin–porphyrin interaction in the UV/VIS spectrum. Light-scattering measurements of the polymer at optimum ratios of PVP to PTS showed, however, that only one porphyrin per four PVP was bound, although the binding constant was extremely high[99].

5.9 Biopolymer Fibres

The most important assemblies of biopolymer fibres are presumably DNA[100,101], sickle cell haemoglobin[102], collagen[103], elastin[104], actin[103], glycoproteins[105] and polysaccharides[106]. They have much in common with non-

[98] S.J. Milder, L. Ding, G. Etemad-Moghadam, B. Meunier, N. Paillous, *J. Chem. Soc., Chem. Commun.*, **1990**, 1131
[99] F.M. El Torki, P.J. Casano, W.F. Reed, R.H. Schmehl, *J. Phys. Chem.*, **1987**, *91*, 3686
[100] W. Saenger, Principles of Nucleic Acid Structure, Springer, New York, **1984**
[101] C.R. Cantor, P.R. Schimmel, Biophysical Chemistry, Part 1, W.H. Freeman, New York, **1982**
[102] R.E. Dickerson, I. Geis, Hemoglobin, Benjamin, Menlo Park, **1983**
[103] T.A. Creighton, Proteins, W.H. Freeman, New York, **1993**
[104] D.W. Urry, *Angew. Chem.*, **1993**, *105*, 859; *Angew. Chem., Int. Ed. Engl.*, **1993**, *32*, 819
[105] M.I. Horowitz, The Glycoconjugates, Vol. 1–4, Academic Press, New York, **1982**
[106] M. Yalpani, Polysaccharides, Elsevier, Amsterdam, **1988**

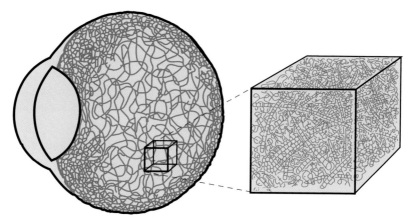

Figure 5.29 *Model of the human eye lens. The fibres consist mainly of collagen and anionic polysaccharides.*

covalent fibre assemblies. The subject is, however, altogether too broad to be covered here. Our contribution handles at first an impressionistic view of the clear gel from which the lens of the eye is made (Figure 5.29). This gel's stability and transparency clearly depends on the fact that chiral polymer fibres have a slight tendency to crystallize out, or in this case, to form cloudy cataracts.

Furthermore, we mention several recent results on synthetic peptides, polysaccharides and nucleic acids. The original papers on this topic can serve as a starting point. Starch of low molecular weight (dextrins) does not entrap iodine and occurs as a linear edge amphiphile. The hydroxy groups are all disposed on one side or the face of the 1a,4e-glucopyranose chain and the hydrogens are disposed on the other. A particularly characteristic property of such an edge amphiphile is its ability to enhance the solubility of lipophilic compounds (steroids) in water. Dextrin definitely increases the solubility of certain steroid hormones by a factor of approximately ten. On the other hand, high molecular weight dextran, which is helical, lowers the solubility of steroids[107].

A fascinating self-assembly process is based on the **spontaneous threading of cyclodextrins or crown ethers onto polyether**[108,109], **polyamine**[110] **or polyurethane**[111] **chains** (Figure 5.30). The formed polyrotaxanes possess several cyclodextrin units which are trapped after appropriate capping of the ends. Polymerization of the threaded cyclodextrins and removal of the central thread then leads to molecular cyclodextrane tubules with an inner hydrophobic tunnel and a diameter of about 5 Å.

In chloroform, the hydrophobic Boc-Val-Val-Ile-Ome tripeptide assembles

[107] D. Balasubramanian, B. Raman, C.S. Sundari, *J. Am. Chem. Soc.*, **1993**, *115*, 74
[108] A. Haradi, J. Li, M. Kamachi, *Nature (London)*, **1992**, *356*, 325
[109] A. Haradi, J. Li, M. Kamachi, *Nature (London)*, **1993**, *364*, 516
[110] G. Wenz, B. Keller, *Angew. Chem.*, **1992**, *104*, 201; *Angew. Chem., Int. Ed. Engl.*, **1992**, *31*, 197
[111] G. Wenz, E. von der Bey, L. Schmidt, *Angew. Chem.*, **1992**, *104*, 758; *Angew. Chem., Int. Ed. Engl.*, **1992**, *31*, 783

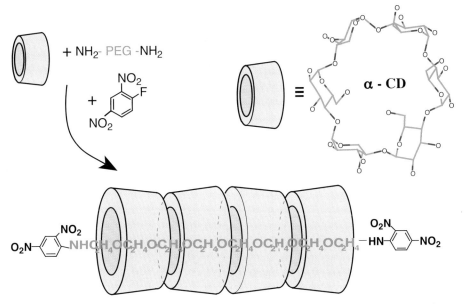

Figure 5.30 *A molecular necklace prepared by synkinesis of poly(ethylene glycol) bis-amine and α-cyclodextrin. The molecular assembly was finally locked by capping with 2,4-dinitrofluorobenzene.*

to micelles. A β-sheet structure was determined by NMR[112]. The model peptide dynorphin A 1–17 (YGG(3)FLRRIR(9)PKLKWDNQ) was analysed via high-resolution NMR in water and methanol. No α-helix was observed. However, NOESY studies of the same peptide dissolved in predeuterated lipid micelles made of dodecylphosphocholine showed that it is helical from residues G3 to R9. The anisotropic nature of the micelle and its small size obviously help to stabilize the peptide in more or less one conformation as opposed to an assembly of conformations in solution[113]. A parallel four-helix bundle was obtained by ruthenium ion-assisted aggregation of a polypeptide equipped with a pyridyl functionality at the *N*-terminus[114]. An amphiphilic peptide with eight leucines in the hydrophobic phase formed a quadruple helix instead of the usual six-form hexameric helical bundles[115]. Such bundles entrap fluorescent probes within their hydrophobic cavity.

The cyclic peptide *cyclo*-[(D-Ala-Glu-D-Ala-Gln)₂] with an even number of alternating D- and L-amino acids **adopted a low-energy, ring-shaped flat conformation** in which all backbone amide functionalities lay almost perpendicularly to the plane of the structure. This synkinon allows subunits to stack in an

---

[112] R. Jayakumar, A.B. Mandal, P.T. Manoharan, *J. Chem. Soc., Chem. Commun.*, **1993**, 853
[113] D.A. Kallick, *J. Am. Chem. Soc.*, **1993**, *115*, 9317
 Amino acid symbols: A Ala; B Asx; C Cys; D Asp; E Glu; F Phe; G Gly; H His; I Ile; K Lys; L Leu; M Met; N Asn; P Pro; Q Gln; R Arg; S Ser; T Thr; V Val; W Trp; Y Tyr
[114] M.R. Ghadiri, C. Soares, C. Choi, *J. Am. Chem. Soc.*, **1992**, *114*, 4000
[115] T.-M. Chin, K.D. Berndt, N.C. Yang, *J. Am. Chem. Soc.*, **1992**, *114*, 2279

antiparallel fashion and participate in backbone–backbone intermolecular hydrogen bonding to produce a contiguous β-sheet structure. Moreover, the alternating D- and L-sequence pushed sidechains out of the middle, thereby creating the desired hollow centre (Figure 5.31). Controlled acidification of the glutamate side chains in alkaline solutions of the *cyclo*-octapeptide triggered the precipitation of tubules hundreds of nanometres long. The internal diameter in the single strands should be as in the gluconamide fibres 7–8 Å, corresponding to the diameter of the cyclopeptide. The observed tubules contained about one hundred parallel strands[116a].

Patch-clamp experiments with the tubules of the similar octapeptide *cyclo*-[(Trp-D-Leu)$_3$-Gln-D-Leu] in phospholipid bilayers indicate short-lived

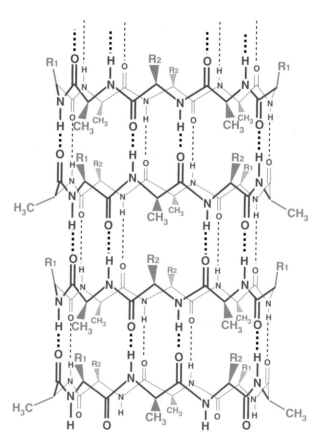

peptide: cyclo[-(D-Ala-Glu-D-Ala-Gln)$_2$-]
$R_1 = CH_2CH_2COOH$
$R_2 = CH_2CH_2CONH_2$

Figure 5.31 *Fibrous assembly of a planar* cyclo-*peptide in water*[116a].

[116] a) M.R. Ghadiri, J.R. Granja, R.A. Milligan, D.E. McRee, N. Khazanovich, *Nature (London)*, **1993**, *366*, 324

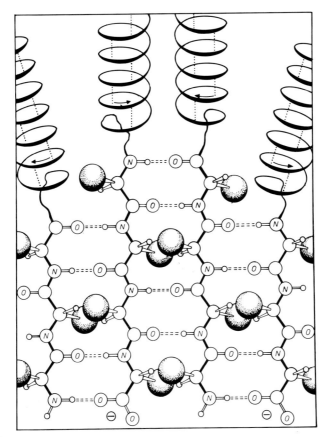

Figure 5.32   *Poly (D-lysine) and (L-lysine) form helices in water above pH = 10. Mixture of the M- and P-helices leads to rapid precipitation of a β-sheet structure.*

potassium and sodium ion channels. The mechanism of rapid tubule assembly–disassembly processes remain, however, obscure. There should be no simultaneous rupture of multiple amide hydrogen bonds at room temperature. As long as the channels cannot be reversibly closed with stoppers, one must consider ion transport along the border line between the lipids and the stiff peptide as a much more likely transport mechanism than flow of ions through the inner tubule (compare Figures 4.18, 4.24 and 4.25)[116b].

The chiral bilayer effect (see Figure 5.15) was also demonstrated in polypeptides, but not in nucleic acids. D- and L-polylysine formed perfect solutions of helical strands at pH 9, but precipitated within a few minutes as pleated sheets if mixed in a 1:1 ratio. Chiral purity is again a pre-supposition for the stability of helices (Figure 5.32).

The chemically synthesized L-d(CGCGCG) was co-crystallized with equimolar D-enantiomer. The enantiomers of nucleic acid helices, e.g.

[116]   b) M.R. Ghadiri, J.R. Granja, L.K. Buehler, *Nature (London)*, **1994**, *369*, 301

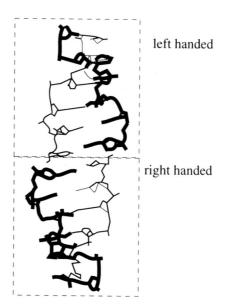

Figure 5.33 *X-ray structure of racemic DNA d(CGCGCG). The L-enantiomers form a right-handed duplex (bottom box), the D-enantiomers a left-handed duplex (top box).*

D-d(CGCGCG) and L-d(CGCGCG), pair only with complementary strands which have the same chirality. Racemic DNA molecules only interact at the end of the strand to form **crystals in which the D-configured double helix's ends stick to two L-configured double helices and vice versa**[117]. In the crystal, a pseudohelix is thus formed which contains half-a-turn of right-handed and half-a-turn of left-handed enantiomer structures (Figure 5.33). There is **no** formation of **polyphosphate sheets**.

5.10 Helical Metal Complexes

The first double-helical metal complex described was the tetrahedral bis(zinc formylbiliverdinate) (Figure 5.34). The helical dimer forms spontaneously, if the planar monomeric zinc complex, which carries an axial water ligand, is dehydrated by HCl[118].

Longer helicates with a central metal ion "wire" can be obtained from oligomeric bipyridines and similar ligands. Oligo-1,2-bipyridines have the potential to behave as bipyridine or terpyridine derivatives. These acyclic hosts undergo self-organization in the presence of metals with tetrahedral or octahedral coordination numbers (Figure 5.35). Double helices with several metal ions in close contact are thus formed[119].

[117] M. Doi, M. Inoue, K. Tomoo, T. Ishida, Y. Ueda, M. Akagi, H. Urata, *J. Am. Chem. Soc.*, **1993**, *115*, 10432
[118] G. Struckmeier, Y. Thewalt, J.-H. Fuhrhop, *J. Am. Chem. Soc.*, **1976**, *98*, 278
[119] K.T. Potts, K.A.G. Raiford, M. Keshavarz-K, *J. Am. Chem. Soc.*, **1993**, *115*, 2793

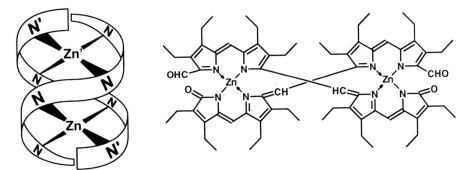

Figure 5.34 *Schematic X-ray structure of the bis-helical zinc octaethylformylbiliverdinate. Dimerization is caused by the tendency of the zinc ion to arrange nitrogen ligands as a tetrahedral ligand field. This cannot be achieved here with a single tetrapyrrole ligand.*

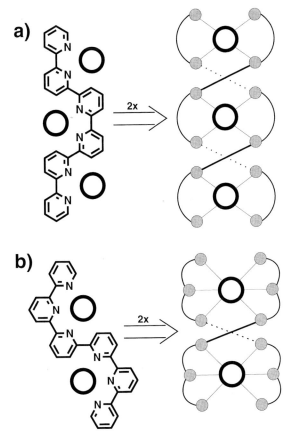

Figure 5.35 *Anticipated tetrahedral (a) and octahedral (b) coordination of a transition metal ion by a sexipyridine.*

The pyridine units, however, must be substituted in order to produce water-soluble oligomers. Oligothymidines formed bis-helices upon titration with Cu^I salts. This arrangement (Figure 5.36) is called an "inverted DNA", because it contains internal (positive) charges and external nucleic acid bases[120]. Cooperative effects normally led to two identical strands in these helicates when copper(I) ions were added to a mixture of two different homooligomers. A micellar helix in chloroform made from corresponding copper(I) oligopyridines with alkyl side chains was also reported. The hydrophobic effect provided the "sticky ends' necessary for the assembly of a multitude of ligands[121]. Lanthanide *triple* helices are also known[122] whereby three ligands were wrapped around the helical axis defined by europium(III) ions. Other copper complexes formed short double helices in solutions, but in their crystals the units were

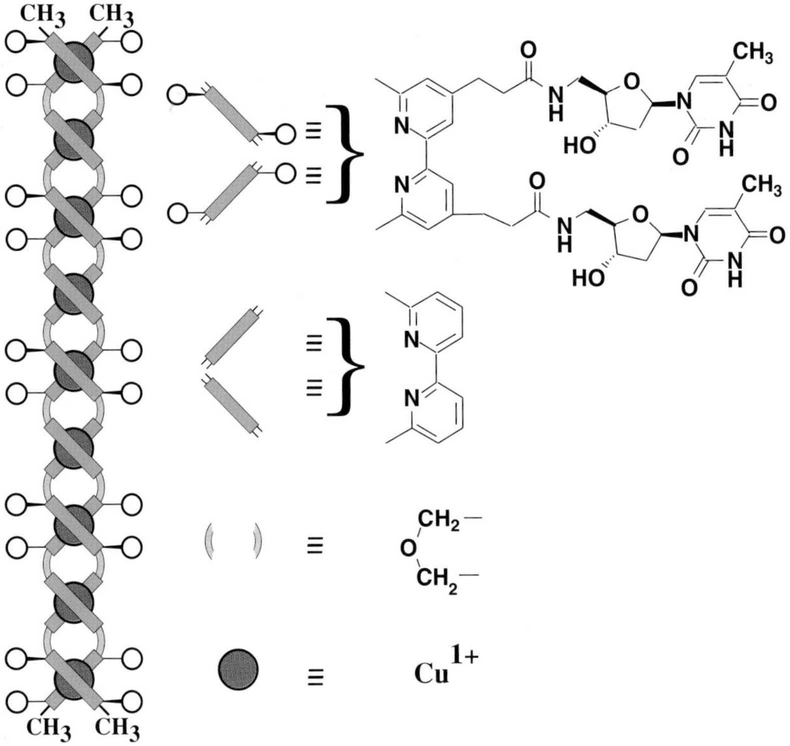

Figure 5.36 *Double helix with "sticky" peripheral nucleic acid base substituents and a central copper(I) ion "wire".*

---

[120] U. Koert, M.M. Harding, J.-M. Lehn, *Nature (London)*, **1990**, *346*, 339
[121] T. Gulik-Krzywioki, C. Fouquey, J.-M. Lehn, *Proc. Natl. Acad. Sci. USA*, **1993**, *90*, 163
[122] C. Piquet, J.-C.G. Bünzli, G. Bernardinelli, G. Hopfgartner, A.F. Williams, *J. Am. Chem. Soc.*, **1993**, *115*, 8197

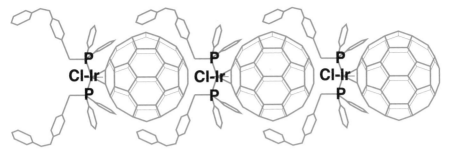

Figure 5.37 *A metalloorganic fullerene wire connected by π–π interactions.*

perfectly aligned and an endless double helix appeared[123]. A similar silver–cobalt complex was also obtained[124].

Vaska-type iridium compounds Ir(CO)Cl(PR$_3$)$_2$ bind readily and reversibly to fullerenes via the iridium atom. Flexible, aromatic substituents on both phosphorus atoms then may accommodate π–π interactions with the convex exterior of C$_{60}$. As a result linear fullerene chains precipitate as black crystallites[125] (Figure 5.37).

5.11 A Stereoselective Pitch Depression in a Cholesteric Phase

With the exception of cholesteric phases, liquid crystals have *not* been taken into consideration in this book because the molecular structure of these giant assemblies is too complex and the study on the molecular level has only been minimal. Another exception concerns a case where the overall pitch in cholesteric phases may be extremely uniform and well-defined. Its changes can even be controlled and monitored by external reagents. When chiral ammonium ions (amino acid esters) are added to a ternary blend of cholesterol chloride, cholesterol nonaoate and a steroidal crown ether, the helical pitch of the mixed cholesteric crystal is altered, resulting in a visible colour change. In the most interesting system of all, the cholesterin was linked to phenyl boronoic acid **19** via a rigid amide group. There is the possibility that in the liquid crystal one monosaccharide molecule may or may not connect two units of **19** to a glucose molecule. For example, one boronic acid binds with the 1,2-diol unit to form a five-membered ring and the other with the 4,6-diol to yield a six-membered ring. Models show that with a given helical sense, this 2:1 complex should have the five-membered boronic ester ring down in the D-glucose case and up with the

---

[123] R.F. Carina, G. Bernardinelli, A.F. Williams, *Angew. Chem.*, **1993**, *105*, 1483; *Angew. Chem., Int. Ed. Engl.*, **1993**, *32*, 1463

[124] E.C. Constable, A.J. Edwards, P.R. Raithby, J.V. Walker, *Angew. Chem.*, **1993**, *105*, 1486; *Angew. Chem., Int. Ed. Engl.*, **1993**, *32*, 1465

[125] A.L. Balch, V.J. Catalano, J.W. Lee, M.M. Olmstad, *J. Am. Chem. Soc.*, **1992**, *114*, 5455

L-glucose enantiomer (Figure 5.38). Correspondingly, the green cholesteric liquid crystalline phase becomes red upon addition of D-glucose and blue upon addition of L-glucose. Diasteromeric hexoses gave similar effects[126].

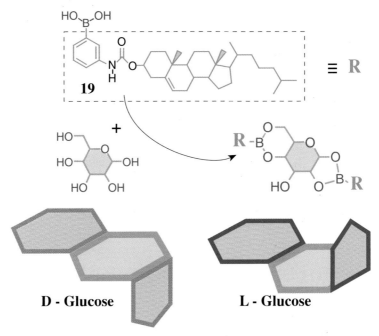

Figure 5.38 *Scheme of the stereoselective recognition of carbohydrates by helical cholesteric phases containing boron esters.*

---

[126] T.D. James, T. Harada, S. Shinkai, *J. Chem. Soc., Chem. Commun.*, **1993**, 857

CHAPTER 6

Molecular Recognition and Nanopores in Surface Monolayers

6.1 Introduction

Molecular monolayers on water (Langmuir films) or on the surface of solid materials (= Langmuir–Blodgett films = LB films) are easy to prepare using commercial Langmuir troughs or similar equipment. Their preparation, analysis and material properties have been so comprehensively covered in two books by Gaines and Ulman[1] that there is no requirement to handle this subject again. This is even more obvious when one realizes that monolayers with dimensions of several square centimetres can neither be considered as "molecular assemblies" nor as "membranes"; rather they constitute a separate liquid phase or solid phase with all the irregularities typical for bulk phases, e.g. statistical formation of domains, microcrystals, holes etc. In principle, one can only prepare and analyse one monolayer at a time and then, each individual monolayer is different from the others, whereas in micelles, vesicles and fibres, one can usually handle a large number of individual compositions on an average basis.

On a re-thought, however, molecular monolayers open unique possibilities for studying chemo- and stereoselective interactions between different amphiphiles and between amphiphiles and signal molecules in aqueous solution. These studies can either be carried out via measurements of molecular surface areas on water or by electrochemical detection of opened and closed nanopores on gold surfaces. Both aspects have not been covered in Ulman's recent book and the words "chiral recognition", "pores or channels", which are of central importance to organic supramolecular chemistry, do not appear at all. The following short paragraphs attempt to deal exclusively with **selective binding interactions** of molecular monolayers. For methodical and technical details, as well as for polymerization reactions within monolayers we refer the reader to Ulman's book.

---

[1] a) A. Ulman, Ultrathin Organic Films, Academic Press, Boston, **1991**
    b) G.L. Gaines, Insoluble Monolayers at Liquid–Gas Interfaces, Interscience, New York, **1966**

6.2 Molecular Recognition at Monolayers on the Water Surface

The olfactory system of humans and animals can discriminate between enantiomeric odorants. L-DPPC monolayers on water are also capable of the same, although their chiral centre is hidden inside the carbon tails. R-Carvone, for example, is found in spearmint and S-carvone in cumin. L-DPPC monolayers spread on subphases containing 5 mM S- and R-carvone **1** and **2** respectively, showing that monolayers on R-carvone are more expanded. R-Carvone molecules obviously have a stereoselective advantage over the S-enantiomer when interacting with L-DPPC monolayers; an effect also seen in thermodynamic data (ΔG, ΔS, ΔH), extracted from temperature dependent surface pressure–area isotherms. They also showed that R-carvone binds tighter to the L-DPPC **3** surface layer than S-carvone. These results are in agreement with the finding that R-carvone is a much stronger odorant to humans than S-carvone on a threshold basis. One could therefore conclude that the stereochemistry of the phospholipid component of olfactory receptor membranes has as much significance as the stereochemistry of the corresponding membrane proteins[2].

S(+)-Carvone R(−)-Carvone   L-DPPC ≡ S-DPPC
     1   **2**   **3**

The chiral discrimination between chiral molecules forming an insoluble Langmuir monolayer has been investigated theoretically[3]. A simple model for a particular chiral amphiphile assumes one chiral carbon which is bound to four different groups (A, B, C and D). A, B and C are restricted to lying on the water/air interface; D is a hydrophobic chain pointing into the air (Figure 6.1a). A, B and C should be arranged clockwise in the S-enantiomer. If A and B are oppositely charged (or are proton donors (+) and acceptors (−)) and the third group is apolar (0), the homochiral arrangement is favoured (Figure 6.1b) and the spontaneous resolution (= separation of enantiomers in domains) may occur. If the preferred interactions occur between like-groups, racemic pair formation (Figure 6.1c) is preferred. No experimental data are available at the moment which support this two-dimensional model, but it should be right.

Chiral discrimination effects, however, have been demonstrated with amphiphiles containing relatively apolar methyl ester and hydroxy head groups. Comparisons of mixed monolayers and crystals consisting of long-chain

---

[2] S. Pathirana, W.C. Weely, L.J. Myers, V. Vodyanoy, *J. Am. Chem. Soc.*, **1992**, *114*, 1404
[3] D. Andelman, *J. Am. Chem. Soc.*, **1989**, *111*, 6536

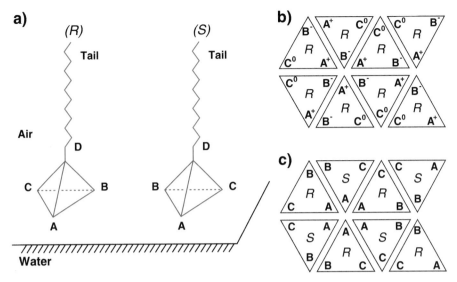

Figure 6.1 *a) Model of the two stereomeric arrangements D (or R) and L (or S) of an amphiphile on a water surface. The groups A, B and C lie counterclockwise for D (R), if one substituent is H.*
b) Chiral discrimination may be homochiral, if the substituents provide strong electrostatic attraction. Domain formation of enantiomers is favoured.
c) Racemate formation or heterochiral discrimination should occur, if interaction between like groups is strongly preferred.

secondary amides with serine, threonine, cysteine and cystine methyl ester head groups indicate that the chiral centre α to the ester group determines chiral recognition[4]. As an example, the surface pressure vs area isotherms for compression and expansion of enantiomeric and racemic stearoyl serine methyl esters **4** at different temperatures are depicted in Figure 6.2a. The differences between the enantiomers only became apparent in collapsed, metastable monolayer states. The phase behaviour involves a transition from the liquid-expanded to the liquid-condensed state, which appears at ~20°C for the racemic films and at ~30°C for the enantiomeric films. **Chiral recognition decreases strongly with increasing temperature**. Epifluorescence micrographs[5] of the racemic film at 0–4 mN/m show domains without long-range order, whereas films cast from enantiomeric material consist of a single, solid domain. Close inspection of the enantiomeric crystal showed striations. Scanning tunnelling electron (= STE) micrographs also only exhibited (2D?) crystals for the enantiomer (Figure 6.2c). **All the investigated cases portrayed a considerable quasicrystalline order in the enantiomeric films in comparison to highly fluid racemic ones.**

---

[4] N.G. Harvey, D. Mirajovsky, P.L. Rose, R. Verbiar, E.M. Arnett, *J. Am. Chem. Soc.*, **1989**, *111*, 1115
[5] J.G. Heath, E.M. Arnett, *J. Am. Chem. Soc.*, **1992**, *114*, 4500

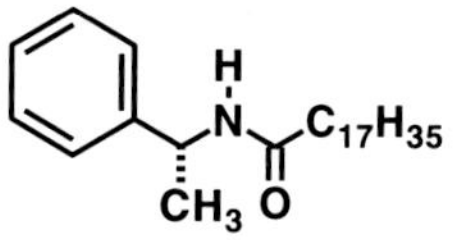

5

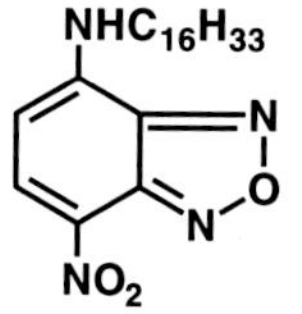

6

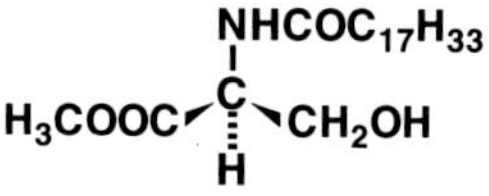

S-SSME
4a

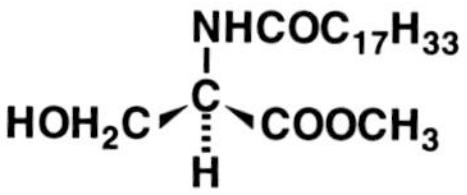

R-SSME
4b

Chiral discrimination effects within surface monolayers may also be employed for separating a racemic surface monolayer into domains of uniform chirality which occurs if the $S{:}S$ or $R{:}R$ interaction is more favourable than the $S{:}R$ interaction (homochiral discrimination)[6]. Such a two-dimensional resolution was triggered off by the sprinkling of pure enantiomer N-(α-R-methylbenzyl)stearamide **5** crystals on to the corresponding racemic monolayer. A rapid decrease of surface pressure well below the equilibrium surface pressure of the racemate was observed[7,8]. This result implies the deposition of R-configured molecules on the added crystals, leaving a partially resolved film which was composed predominantly of S-molecules.

Direct evidence of chiral monolayers came from fluorescence micrographs[9] of S-stearoylserine **4a** with **6** as a fluorescent probe, which was not dissolved in the 2D crystals. Curved dendrites or small domains are formed, which are curved counterclockwise in L-serine amide monolayers (Figure 6.2d) and clockwise if the head group consists of R-serine (no Figure).

Crystalline head group arrangements of Langmuir monolayers can be used as molecular templates for the oriented nucleation of organic and inorganic crystals. One example is calcium carbonate, which grows from supersaturated calcium bicarbonate solutions in the presence of carbon dioxide[10,11]. In the absence of a surface monolayer, only rhombohedral calcite crystals were formed. In the presence of compressed stearic acid monolayers, only pseudo-hexagonal vaterite crystals were found and these appeared as uniform lens-shaped discs. Obviously the Ca^{2+} counter-ions to the carboxylate head group of the surface monolayer had triggered off the crystallization process, forming the first crystal face, as only vaterite possesses a (001) face consisting purely of

[6] M. Stewart, E.M. Arnett, *Topics in Stereochemistry*, **1982**, *13*, 1
[7] E.M. Arnett, O. Thompson, *J. Am. Chem. Soc.*, **1981**, *103*, 968
[8] E.M. Arnett, J. Chao, B.J. Kinzig, M.V. Stewart, O. Thompson, R.J. Verbiaer, *J. Am. Chem. Soc.*, **1982**, *104*, 389
[9] K.J. Stine, J.Y.-J. Uang, S.D. Dingman, *Langmuir*, **1993**, *9*, 2112
[10] S. Mann, B.R. Heywood, S. Rajara, J.D. Birchall, *Nature (London)*, **1988**, *334*, 692
[11] S. Mann, in S. Mann, J. Webb, R.J.P. Williams (eds), Biomineralization, VCH, Weinheim, **1989**, p. 47ff

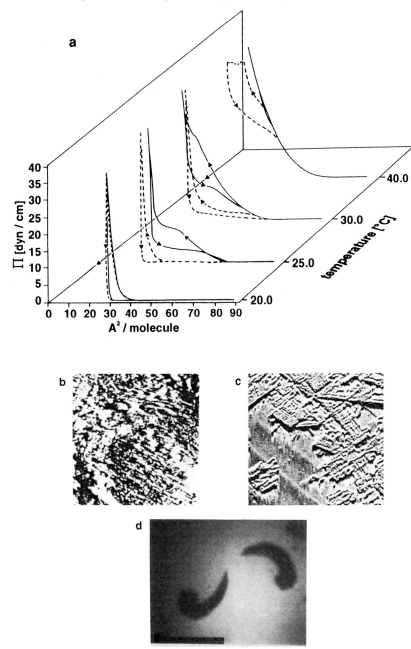

Figure 6.2 *a) Surface pressure vs area isotherms for compression and expansion of enantiomeric (---) and racemic (—) stearoylserine methyl esters on water at various temperatures. b) Scanning tunnelling electron micrograph of collapsed domains in racemic and c) enantiomeric films. Only the latter film is crystalline. d) Fluorescence micrographs of surface monolayers of the L-configured ester containing 0.3 mol% of fluorescent probe. The bar on the lower left represents 125 μm.*

calcium ions. The trigonal carbonate anions then arrange themselves perpendicularly to the calcium face, an action which is equivalent to the orientation of the carboxylate head groups of the stearate monolayer. The stereochemical arrangement of the vaterite structure rather than that of calcite was thus enforced by the sequence of pure calcium and carbonate layers, a motif which was then extended from the first calcium layer throughout the whole vaterite crystal (Figure 6.3). This mechanism does not involve epitaxial matching. The inter-headgroup spacing on the monolayer is about 5 Å compared with a Ca–Ca distance of 4.13 Å in vaterite and 4.96 Å in calcite. If geometric matching was a fundamental property of the system, then calcite would be the favoured polymorph. **Stereochemical and electrostatic matching obviously override structural mismatch at the interface.**

When the planar carbonate motif was replaced by the tetrahedral sulfate motif in the growth of barium sulfate crystals under n-eicosyl sulfate, monolayers were obtained with irregular morphologies, suggesting a negative kinetic rather than a positive structural influence of the organic template on nucleation of $BaSO_4$. The hexagonal lattice of sulfate head groups provides only a limited

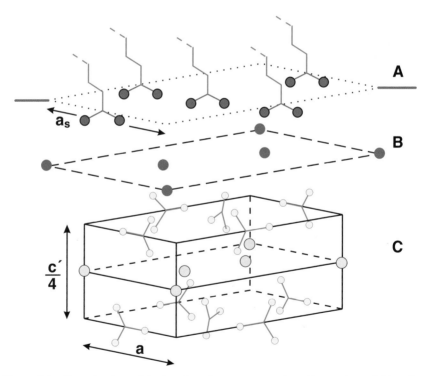

Figure 6.3 *Organization of the interface between a stearic acid monolayer (A), calcium ions (B) and carbonate and calcium layers (C). (C) corresponds to a vaterite (= $CaCO_3$) subcell. There is no geometric matching between the stearate groups (in A; spacing 50 Å) and the Ca–Ca distance in vaterite (C; spacing 4.13 Å). The two-layer sub-unit cell motif of the A–B layer, however, repeats in vaterite (C).*

geometric match to BaSO$_4$ crystals. A complex dendritic morphology then results from spatially restricted interfacial growth coupled with rapid growth into the bulk solution[12]. Soluble phosphonates are known to be potent inhibitors of barium sulfate crystallization. Nevertheless, phosphonate monolayers induced barium sulfate precipitation at the air/water interface. Geometric and stereochemical complementarity of the organic–inorganic interface were again key factors in the formation of the unique morphologies[13] which appeared.

Monomolecular films formed from pure enantiomers of amino acids initiate the rapid nucleation of oriented glycine crystals when structural information in the monolayer was correlated with specific lattice parameters in the nascent crystal[14]. Surface monolayers displaying small molecular cross-sectional areas in the range of 25–29 Å^2, e.g. palmitoyl lysine or stearoyl aspartate, caused fast nucleation of the glycine crystals. When monolayers of R configuration were used, all the glycine crystals (010) were oriented and occluded N'-(2,4-dinitrophenyl)-S-lysine whereby yellow plates appeared. On the other hand, the usual colourless pyramids crystallized when S monolayers were used. **Morphology, chirality and the ability to occlude guest molecules are thus completely determined by the chirality of the monolayer and not by the achiral molecules which form the crystal.**

The head groups of surface monolayers can also be used as chemo- and stereoselective receptors for small molecules in the subphase. Alkylated azacrowns, for example, form highly oriented monolayers on water. They hardly discriminate between AMP and ATP in the subphase if it is buffered, but ATP binding action is stronger than that of the other two in distilled water[15]. Cationic guanidinium head groups bind to the phosphate groups of ATP and other mononucleotides with binding constants in the order of 10^7 M^{-1} (Figure 6.4a)[16]. If the guanidinium amphiphile is mixed with an adenine component in a 1:1 ratio, then UMP binding (Figure 6.4b) is more suitable in comparison to AMP binding[17]. Diazo dyes in the side-chain have been used to detect domain formation by UV spectroscopy.

Convergent carboxylic acids of long-chain derivatives of Kemps's acid **7** are useful receptors for fitting nitrogen heterocycles, e.g. benzimidazole[18] and is one of very few examples for the selective binding of electroneutral molecules in water.

Selective binding between polymeric substrates and monolayered receptors has been accomplished more frequently. One of the most important recognition processes in nature is the pairing of nucleic acid bases, which can be mimicked with Langmuir films. Lipid monolayers with adenine head groups expand upon

---

[12] B.R. Heywood, S. Mann, *J. Am. Chem. Soc.*, **1992**, *114*, 4681
[13] R. Heywood, S. Mann, *Langmuir*, **1992**, *8*, 1492
[14] E.M. Landau, S. Grayer Wolf, M. Levanon, L. Leiserowitz, M. Lahav, J. Sagir, *J. Am. Chem. Soc.*, **1989**, *111*, 1437
[15] C. Merksdorf, T. Plesnivy, H. Ringsdorf, P.A. Suci, *Langmuir*, **1992**, *8*, 2531
[16] D.Y. Sasaki, K. Kurihara, T. Kunitake, *J. Am. Chem. Soc.*, **1991**, *113*, 9685
[17] D.Y. Sasaki, K. Kurihara, T. Kunitake, *J. Am. Chem. Soc.*, **1992**, *114*, 10994
[18] Y. Ikeura, K. Kurihara, T. Kunitake, *J. Am. Chem. Soc.*, **1991**, *113*, 7343

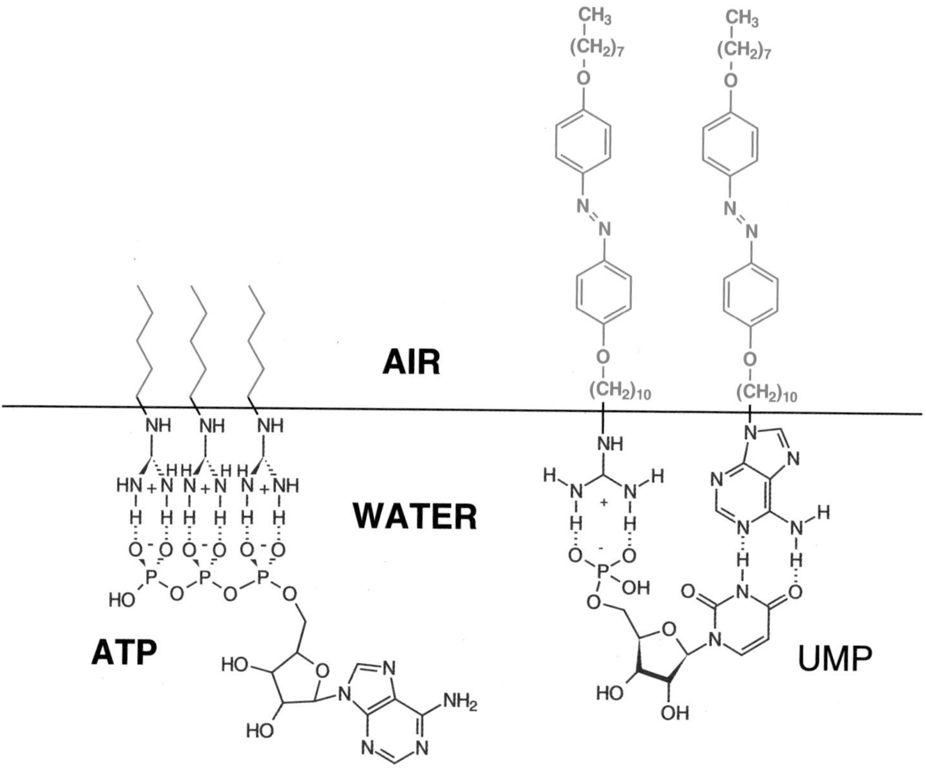

Figure 6.4 *a) ATP binds in a ratio 1:3 to guanidinium head groups of surface monolayers.*
b) UMP fits better than AMP[15-17].

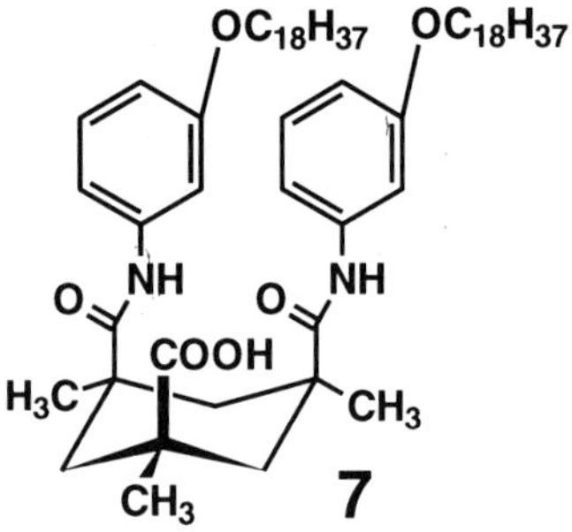

addition of complementary polynucleotides, e.g. poly(U) to the subphase. Non-complementary polynucleotides, e.g. poly(A) have less effect. The finding that the poly(U) effect also remains in the solid-analogue phase, whereas poly(A) is completely rejected upon solidification of **8** is most remarkable[19].

Monolayers of lipid **9** with a biotin head group bind strongly to strepta-

---

[19] H. Kitano, H. Ringsdorf, *Bull. Chem. Soc. Jpn*, **1985**, *58*, 2826

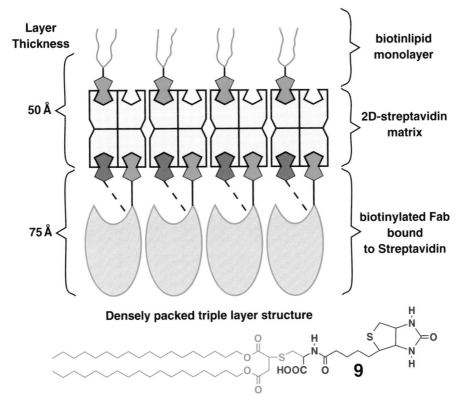

vidin dissolved in the subphase; crystalline domains are thus formed. Since streptavidin has four binding sites for biotin it can again serve as a template for the binding of biotinylated FAB[20]. Stable triple layers form on the surface (Figure 6.5).

Figure 6.5 *Biotin head groups of a surface monolayer bind strongly to the protein streptavidin containing four biotin binding centres. A crystalline protein layer can now be observed by electron microscopy. The other binding centres can now be used to form a third monolayer with biotinylated proteins, e.g. the FAB fragment of F immunoglobulins*[20].

[20] J.N. Herron, W. Müller, M. Pandler, H. Riegler, H. Ringsdorf, P.A. Suci, *Langmuir*, **1992**, *8*, 1413

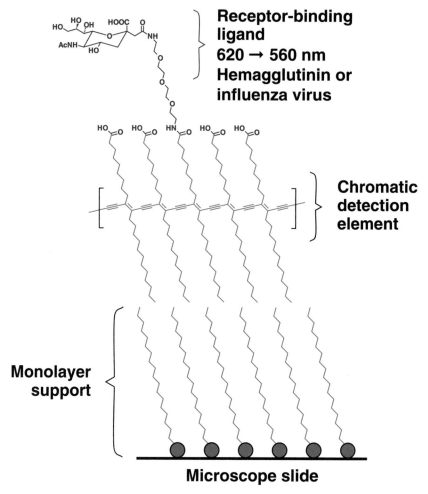

Figure 6.6 *Statistically distributed sialic acid on the polar surface of polymeric monolayers is recognized by the viral lectin called hemagglutinin (compare with Con A, Figures 4.11 and 4.12). A protein surface layer is formed, which disturbs the order in the polymeric monolayer below. Colorimetric measurements of the colour change from blue to red allow quantitative determination of the protein concentration.*

Diacetylenic lipid monomers, such as the long-chain carboxylates and corresponding sialic acid derivatives are rapidly polymerized in monolayers to form a conjugated diacetylene backbone of alternating ene-yne groups (Figure 6.6). The monolayers change colour from blue to red with an increase in temperature or other disordering exterior influences. The polymeric monolayer adsorbed on a smooth support also displays this colour change upon addition of the viral lectin, haemagglutinin. The colorimetric response of the film is directly proportional to the added quantity of haemagglutinin or influenza virus. Addition of competitive inhibitors suppressed the response[21].

6.3 Molecular Recognition at LB Films and Self-Assembled Monolayers

Langmuir–Blodgett films (LB films) are molecular mono- or multilayers, which have been transferred from the water–air interface onto a solid substrate and can be removed with organic solvents or aqueous detergent solutions. On the other hand, self-assembled monolayers (SAM) form spontaneously on the immersion of a solid into the organic solution of a reactive surfactant. SAMs are bound by chemisorption and cannot be removed with organic solvents or detergents[22]. Both LB films and SAMs can be considered as being highly ordered, closely packed molecular assemblies with several defects. Without going into great detail, we describe the properties of a few characteristic SAMs, starting with the classical silicate subphase which forms stable esters with activated fatty acid derivatives. We then proceed with the most popular of the alkanethiol monolayers on metals, particularly gold.

Fatty acid monolayers on water surfaces have sharp melting points, e.g. 278 K for myristic acid (C_{14}), 301 K for palmitic acid (C_{16}) and 317 K for stearic acid (C_{18})[22]. Below the melting point, synchrotron X-ray diffraction, electron microscopy and infrared spectroscopy on a SiO_2 substrate ($\delta(CH_2)_{sym} = 2917.1$ cm^{-1}) clearly showed that the myristic acid monolayer is crystalline even at very low surface pressures[23,24]. At room temperature, it produces amorphous monolayers at low and high surface pressures as previously shown by electron microscopy with gold decoration[22] and by infrared spectroscopy ($\delta(CH_2)_{sym} = 2923.8$ cm^{-1}). **Through an increase in surface pressure, amorphous monolayers produced a plateau, but not crystalline monolayers**[25] (Figure 6.7). **This could indicate a conformation conversion from *gauche* to *trans* in the amorphous layers, a change which does not occur in crystalline layers.** It was also shown by pH-dependent electron microscopy that anionic arachidonate monolayers shifted from an amorphous to a crystalline state upon surface compression. The free acid monolayers were always crystalline.

We compare these monolayers with SAMs on gold. Features of the alkanethiol monolayers on gold which make them particularly attractive include their relatively uncomplicated preparation, their thermodynamic and chemical stability and mechanical strength. The stability transpires through the formation of a chemical Au—S bond which gives rise to two-dimensional lattice structure in which the head groups form an overlayer with a lattice constant of

---

[21] D.A. Charych, J.O. Nagy, W. Spevak, M.D. Bednarski, *Science*, **1993**, *261*, 585
[22] T. Kajiyama, Y. Oishi, M. Uchida, Y. Tanimoto, H. Kozuru, *Langmuir*, **1992**, *8*, 1563
[23] R. Putta, J.B. Peng, B. Lin, J.B. Ketterson, M. Prakash, *Phys. Rev. Lett.*, **1987**, *58*, 2228
[24] K. Kjaer, J. Als-Nielsen, C.A. Helm, P. Tippman-Kryer, H. Möhwald, *J. Phys. Chem.*, **1989**, *93*, 3200
[25] T. Kajiyama, Y. Oishi, M. Uchida, Y. Takashima, *Langmuir*, **1993**, *9*, 1978

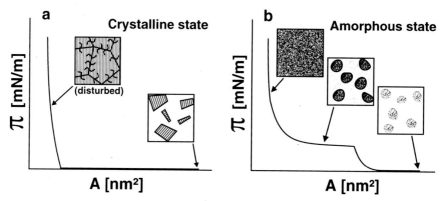

Figure 6.7 *Surface monolayers may be crystalline (see also Figure 7.1) or amorphous. The transition from crystalline to amorphous can be triggered, for example, by (i) a rise in temperature of the subphase leading to a melting process or, (ii) by a change of pH in the subphase, leading to a charged monolayer with non-directed repulsion or, (iii) by adding the other enantiomer to a crystalline pure enantiomeric monolayer (see Figures 6.2 and 7.1). Amorphous monolayers often produce a plateau, indicating conformational or inclination changes, which are less likely in 2D crystals.*

4.99 Å and a surface area of 21.4 Å^2 per chain. The ordering of the head groups persists over regions which are at least a few hundred angstrøms in length. On the outer surface, however, true long range order is absent and the hexagonal ordering of terminal methyl groups is confined to domains with linear dimensions of tens of angstrøms. A 224-molecule Monte Carlo simulation is characterized by well-differentiated domains, indicating that the monolayer surface is extensively pitted[26]. Membrane thickness varied from 18 to 23 Å. Earlier calculations with smaller numbers of molecules did not reveal such fluctuations and were therefore not in agreement with experimental results.

Other surface monolayers containing $CH_3(CF_2)_7NHCO(CH_2)_2SH$ and $C_{12}H_{25}SH$ were reported as being much smoother. In this case, amide hydrogen bond chains may enforce long range order[23]. Octadecanethiolate on gold produced atomically resolved images with atomic force microscopy (AFM). They exhibited a periodic hexagonal pattern with nearest and next-nearest neighbour distances of $a = 0.52$ and $b = 0.90$ nm, corresponding well with the same distances of the (111) face of gold (0.29 and 0.50 nm)[28].

Self-assembled monolayers (SAMs) of long-chain alkylthiols on gold, silver and copper[29] (bound as surface thiolates) and of fatty acids (bound as carboxylates to surface oxides) on silver, copper and aluminium[30] produced well-

[26] J.I. Siepmann, I.R. McDonald, *Langmuir*, **1993**, *9*, 2351
[27] W. Mizutani, D. Anselmetti, B. Michel, in P.E. Booechl, C. Joachim, A.J. Fisher (eds), Computations for the Nano-Scale, Kluwer, Dordrecht, **1993**, p. 43
[28] C.A. Alves, E.L. Smith, M.D. Porter, *J. Am. Chem. Soc.*, **1992**, *114*, 1222
[29] P.E. Laibnis, G.M. Whitesides, D.L. Allara, Y.T. Tao, A.N. Parikh, R.G. Nuzzo, *J. Am. Chem. Soc.*, **1991**, *113*, 7152
[30] Y.-T. Tao, *J. Am. Chem. Soc.*, **1993**, *115*, 4350

defined infrared spectra, which allowed the determination of the tilt angle and, much more significantly, the conformation. The latter is generally *all-trans* with several *gauche*-conformations near the carboxylate head groups on an aluminium surface[30]. The tilt angle for thiolates is 27° on gold, and 12° on copper and silver. The monolayers on silver are, however, sensitive to air, and also to extended exposure to the solution containing the thiol. Fatty acid monolayers all have tilt angles of between 15° and 25° and show a strong odd–even effect on silver: the intensity of the asymmetric stretch of CH_3 groups occurring around 2965 cm^{-1} alternates between odd and even carbon chains and is higher for an even-numbered carbon chain. The symmetric stretch made around 2877 cm^{-1} alternates in the opposite direction[30] (Figure 6.8). Similar effects were observed for thiols on gold, but not for thiols on silver.

Alternating effects obviously become apparent only on the most stable and best-defined surface monolayers. Tight binding to the substrate is a prerequisite. **Thiols on gold and fatty acids on silver are therefore the choice systems for stereochemical investigations.** Furthermore, CH_2 stretch modes showed a trend to shift to a lower frequency as the chain length increased: i.e. from 2924 and 2854 cm^{-1} for a C-9 acid to 2914 and 2848 cm^{-1} for a C-14 acid. The lower frequencies are identical to those of crystalline bulk samples, indicating crystallinity of the long-chain surface monolayers.

Hexadecanethiol monolayers also self-assemble on mercury surfaces and provide an extremely low defect density[31]. Alkanethiols can likewise be assembled on GaAs (100) surfaces and can act as a useful semiconductor[32]. Alkanenitriles bind side-on ($=\mu^2$) to gold and copper surfaces. The infrared signal in the range of 2000–2500 cm^{-1} is replaced by a 1570–1630 cm^{-1} band which is close to μ^2-coordinated nitrile on platinum. Molecular dynamics calculations for —SH and —SCH_3 on gold produced two chemisorption modes very close in energy[30].

Thiol monolayers are not removed by solvents, but by sulfur-active chemicals which pass through the surface monolayers. Laser desorption mass spectrometry has shown that thiolate molecules are intact on the gold surface, but through air oxidation, some sulfonates develop[35]. The relative stability of alkanethiol SAMs on gold to air oxidation is to be expected due to the covalent nature of the S—Au bond. Photooxidation via UV excitation of electrons in the metal surface is, however, possible and leads to sulfonate salts which have again been characterized by mass spectrometry as well as by XPS[36]. Alkenethiolate monolayers can best be desorbed from gold by a one-electron reductive path[37]. Stable monolayers on gold were also obtained with benzenesulfinate,

---

[31] A. Demoz, D.J. Harrison, *Langmuir*, **1993**, *9*, 1046
[32] C.W. Sheen, J.-X. Shi, J. Martensson, A.N. Parik, D.L. Allara, *J. Am. Chem. Soc.*, **1992**, *114*, 1514
[33] U.B. Steiner, W.R. Caseri, U.W. Suter, *Langmuir*, **1992**, *8*, 2771
[34] H. Sellers, L.A. Ulman, Y. Shnidmann, J.E. Eilers, *J. Am. Chem. Soc.*, **1993**, *115*, 9389
[35] Y. Li, J. Huang, R.T. McIver, J.C. Hemminger, *J. Am. Chem. Soc.*, **1992**, *114*, 2428
[36] J. Huang, J.C. Hemminger, *J. Am. Chem. Soc.*, **1993**, *115*, 3342
[37] D.E. Weisshaar, B.D. Camp, M.D. Porter, *J. Am. Chem. Soc.*, **1992**, *114*, 5806

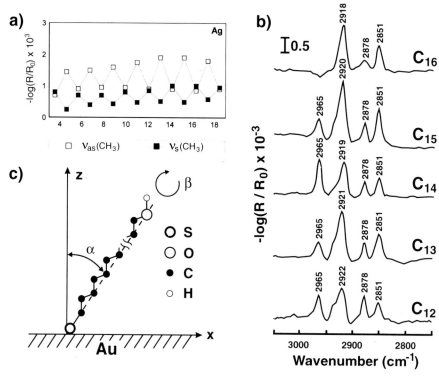

Figure 6.8 *Even–odd effects on solid surfaces as detected by infrared spectroscopy.*
a) IR intensities for methyl stretching modes as a function of chain length for $CH_3(CH_2)_mCOOH$ monolayers on silver. $\delta_{as}(CH_3)$ around 2964 cm^{-1}; $\delta_s(CH_3)$ around 2877 cm^{-1}. δ_a is higher for odd-numbered alkane chains[30].
b) Infrared bands δ_{as} corresponding to CH_2 and CH_3 stretching modes of alkyl thiolates on gold.
c) Schematic side view of an alkyl chain. α is the tilt angle to the surface normal; β defines the rotation of the alkyl chain axis in relation to the x,z plane.

and they are easily displaced by benzenethiol[38]. Attempts to efficiently cleave monolayers on silicate from the subphase were unsuccessful.

Next, we turn to the application of monolayers as selective receptors. The most simple case is the adsorption of multivalent metal ions via complexing head groups. Mixed monolayers of thiolated amphiphiles on gold with acetoacetate head groups have, for example, been employed in the construction of ion-recognizing monolayer membranes[39]. **Only redox-active metal ions can be detected electrochemically**. Tightly bound ions allow electron transfer through the membrane, whereas non-binding ions are denied access to the electrode by the same membrane. The formation of metal ion complexes of the mentioned β-dicarbonyl ligands requires the transformation of the ligand to the enol form

[38] J.E. Chadwick, D.C. Myles, R.L. Garrell, *J. Am. Chem. Soc.*, **1993**, *115*, 10364
[39] I. Rubinstein, S. Steinberg, Y. Rot, A. Shanver, J. Sagiv, *Nature (London)*, **1988**, *332*, 426; **1989**, *337*, 514

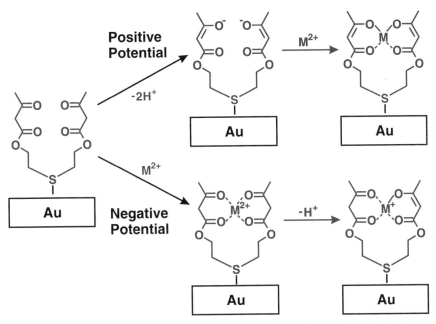

Figure 6.9 *Possible mechanism for the field-driven ionic binding to thiobis(ethyl acetoacetate) moieties arranged in a compact monolayer on gold. It includes a field assisted enolization in acid solution followed immediately by a binding of a metal ion. At negative E the first step may include penetration of the ion into the monolayer to form a weak diketone complex*[40].

(Figure 6.9), thus allowing the fine tuning of organic ligands for selective ion binding through changes of the applied potential[40]. Positive field driven enolization is believed to enhance binding of Cu^{2+} and Pb^{2+} in electroneutral complexes, whereas negative fields stabilize positively charged complexes (Figure 6.9). At zero charge, no metal ions were detected in acid solution.

The electrochemical quartz crystal microbalance (EQCM) can be used to monitor mass changes which occur at electrode surfaces covered with molecular monolayers. The adsorption or desorption of counterions and accompanying solvents in redox processes are responsible for such changes. For example, EQCM was used to characterize the **reversible chloride association** following oxidation of surface-bound ferrocene amphiphiles to the ferricenium state. In this case, the binding to gold was achieved by a terminal ammonium ion, its separation from the electrode being prevented by the long alkyl chain[41]. The membrane coated quartz crystals are, of course, much more sensitive to high molecular weight biopolymers in bulk solution than to smaller ions or molecules. A 10-mer deoxynucleotide with a mercaptopropyl group at the 5'-phosphate end was bound to the gold surface of a 9 MHz quartz. The sequence of the probe DNA was complementary to a natural single stranded 7249 base

[40] S. Steinberg, Y. Tor, E. Sabatini, I. Rubinstein, *J. Am. Chem. Soc.*, **1991**, *113*, 5176
[41] H.C. de Long, J.C. Donohue, D.A. Buttry, *Langmuir*, **1991**, *7*, 2196

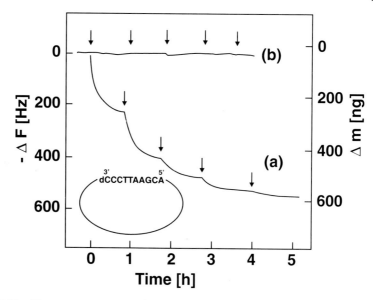

Figure 6.10 *Nanogram quantities of DNA bound to surface monolayers can be measured time-dependently and quantitatively with a 9 MHz quartz for a few dollars.*

pair DNA (MW 2.4×10^6). 120 ng of the latter in 10 mL of water caused a massive frequency decrease of about 200 Hz (Figure 6.10)[42].

On the other hand, long-chain oligo(ethylene oxides) in hydrophobic SAMS reject biopolymers, e.g. proteins[43]. Stabilization of colloids by solvophilic polymers can be explained by a concept called "steric repulsion" which means disfavourable compression and resulting loss of chain mobilities in colliding polymers.

With respect to the distance between electrode surface and redox pair, the redox potential can also be measured with the functionalized alkanethiol monolayers on gold. N,N'-Dialkylbipyridinium salts, $CH_3(CH_2)_n$–$(4,4'\text{Bipy})^{2+}$ –$(CH_2)_m SH \cdot 2Cl^-$, with one terminal SH group and CH_2 groups with $n = 0$, $m = 12$; or $n = 9$, $m = 3$; or $n = 9$, $m = 10$ were bound to gold electrodes on a quartz crystal microbalance[44]. The reduction potential was lowest (-450 mV), where the viologen was close to the water surface and rose by 100 mV (-350 mV) when located within the alkyl chain region. Approximately three water molecules were transported out of the hydrophobic monolayer per anion lost. The viologen chloride on the water surface, however, lost 19 water molecules but no differences appeared when ClO_4^- was used as a counterion[44]. A chemical potential between identical redox pairs which are localized inside, outside or in the centre of the membrane can be thus produced.

[42] Y. Okahata, Y. Matsunobu, K. Ijiro, M. Mukae, A. Murakami, K. Makino, *J. Am. Chem. Soc.*, **1992**, *114*, 8299
[43] K.L. Prime, G.M. Whitesides, *J. Am. Chem. Soc.*, **1993**, *115*, 10714
[44] H.C. De Long, D.A. Buttry, *Langmuir*, **1992**, *8*, 2491

Molecular Recognition and Nanopores in Surface Monolayers

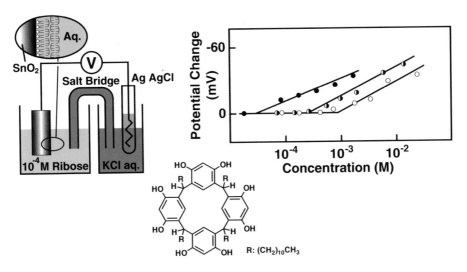

Figure 6.11 *Potentiometry and the response of the calixarene coated SnO_2 electrodes upon the addition of ribose (●), galactose (◐) and glucose (○).*

The most spectacular results, however, were obtained with water-dissolved carbohydrates and a calixarene bound to SnO_2 glass electrode surfaces (Figure 6.11). The adsorption of 10^{-3} molar concentrations of ribose, for example, changed the surface potentials by 30 mV. The detection limit was 2×10^{-5} M with a time response of about 15 minutes. The electrochemical mechanism of the potentiometric response is not established. Nevertheless, the sensitivity of carbohydrate recognition in aqueous solutions appeared as a most surprising and welcome result[47]. In bulk organic solvents such as CCl_4 or $CHCl_3$, two calixarene molecules encapsulate two β-methyl-D-glucopyranoside molecules[45] with a binding constant of around 2×10^8 M^{-4} {K = [complex]/([calixarene][carbohydrate]^4); see reference 45 for details of Hill plot}.

6.4 Nanopores in Self-Assembled Monolayers

Self-assembled monolayers can possess a chiral surface (see previous section) or well-defined holes on a molecular scale ("nanopores"). In the following section we indicate a recent methodology for the synkinesis of nanopores. Their use as receptor sites has as yet been extremely limited. This situation, however, is changing rapidly.

The classical case of surface monolayers are the stationary phases for liquid chromatography. Dimethyloctadecylsilyl-modified silica ($= C_{18}$-silica) is a representative and popular member of the alkylsilyl silicas. From the supra-

---

[45] Y. Kikuchi, Y. Tanaka, S. Sutarto, K. Kobayashi, H. Toi, Y. Aoyama, *J. Am. Chem. Soc.*, **1992**, *114*, 10302
[46] K. Kobayashi, Y. Asahawa, Y. Kato, Y. Aoyama, *J. Am. Chem. Soc.*, **1992**, *114*, 10307
[47] K. Kurihara, K. Ohto, Y. Tanaka, Y. Aoyama, T. Kunitake, *J. Am. Chem. Soc.*, **1991**, *113*, 444

$$\text{—O—Si(CH}_3)_2\text{—(CH}_2)_8\text{-CHD-CHD-(CH}_2)_7\text{-CH}_3$$

a low-loading C18-silica with no added liquids

b low-loading C18-silica with added liquids

c high-loading C18-silica with no added liquids

d high-loading C18-silica with added cyclohexane

e high-loading C18-silica with added water

Figure 6.12 *The mobility of deuterated silane monolayers on silical gel can be studied by solid state deuterium NMR spectroscopy. The degree of coating as well as the environment can be modified. The above models summarize the results. These modified silica surfaces are the most simple models for receptors of apolar molecules. The unique advantage of these probes is the large quantity of monolayer in a small volume.*

molecular chemistry point of view, these phases possess the unique advantage of being produced in quantities which can be studied by ^{2}H-NMR solid state spectroscopy. A "high-loading" C_{18}-silica, for example, contains 195 mg silane/g silica. If only 70 mg/g are applied, a "low-loading" silica sample, which leaves about 70% empty space on the silica surface, or very large "pores" are obtained. The CD_2 groups within the CH_2 chains can then be analysed in dry samples or in the presence of water, cyclohexane or other solvents.

Various results are depicted in Figure 6.12. The spectra of dry, low-loading C_{18}-silica corresponded with highly disordered chains which interacted strongly with the silica surface (Figure 6.12a). The addition of a solvent displaced the hydrocarbon chains from the silica surface and "liquid" C_{18} chains were observed (Figure 6.12b). In this state, it should be possible to entrap hydrophobic compounds within the silica surface monolayer. By addition of liquids (Figure 6.12c) high-loading samples were changed minimally, but unusually well-ordered systems in the centre region were discovered after the addition of cyclohexane (Figure 6.12d). Shape selective surface adsorption via replacement of cyclohexane molecules should be favoured in this case. Low motional rates for high-loading C_{18}-silica with added H_2O suggest aggregation of the C_{18} chains. Since the ^{2}H-NMR results suggest that very few CH_2 groups interact directly with the silica surface, a "haystack" configuration is most likely (Figure 6.12e)[48] and hydrophilic and hydrophobic domains appear side by side. **These pictures not only give a good description of chromatographic behaviour of stationary phases, but can probably also be taken as models for partially covered metal electrodes.**

Self-assembling monolayers with irregular nanopores on the surface of gold electrodes were prepared by admixing isoprenoid alkyl chains with regular alkyl chains. A mixed $C_{18}SH/C_{18}OH$ (7:3) Langmuir film was, for example, transferred onto gold-coated glass electrodes. Its passivating properties against the Ru^{2+}/Ru^{3+} redox couple in water are identical to those of pure $C_{18}SH$-chemisorbed monolayers. If the quinone **10** is incorporated into the passivating $C_{18}SH/C_{18}OH$ monolayer, its passivating character is annulled. The reduction of the quinone occurs at individual sites and it is presumed that methyl groups of the **isoprenoid side-chains provide a gate of sufficiently large diameter to allow $Ru(NH_3)_6^{3+}$ ions to approach the electrode surface**[49].

10

If 4-hydroxythiophenol, $HS(C_6H_4)OH$, is added to the self-assembling $C_{16}SH$ amphiphile, the phenol ring occupies some of the gold surface, but is too thin to passivate the electrode. Defects are thus introduced into the passivating framework of the $C_{16}SH$ monolayer (Figure 6.13), and the ruthenium current can be used as a direct measure of defect density[50]. No reactions of the defects with external substrates have so far been reported. A similar system has, however, been constructed with spironolactone steroid **11** and $C_{16}SH$. Again, the steroid lies flat on the gold surface and blocks binding sites for $C_{16}SH$ but does not passivate the electrode. The $Fe(CN)_6^{3-}/Fe(CN)_6^{4-}$ pair served as a

---

[48] R.C. Zeigler, G.E. Maciel, *J. Am. Chem. Soc.*, **1991**, *113*, 6349
[49] R. Bilewicz, M. Majda, *J. Am. Chem. Soc.*, **1991**, *113*, 5464
[50] O. Chailapaaahul, R.M. Crooks, *Langmuir*, **1993**, *9*, 884

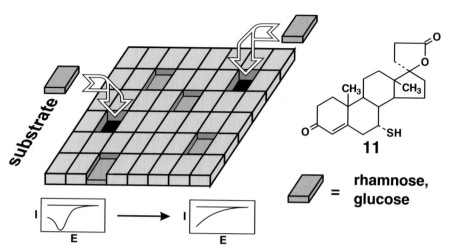

Figure 6.13 *A framework of a covalently bound $C_{16}H_{33}SH$ monolayer (green blocks) is interrupted by thinner, also covalently bound template molecules (black holes), e.g. a quinone- or steroid-thiol. These template molecules serve as gates for electron transfer reactions and can be closed by "signal" molecules (grey blocks) which bind to those templates, e.g. carbohydrates which react with the C-17 substituent of steroid **11**. The current of redox-active ions from the bulk water volume stops.*

current measure for the "steroidal defects" in the $C_{16}SH$ monolayer. If one adds carbohydrates to the aqueous bulk phase, the current drops to much lower values and is not restored if the electrode is washed with distilled water. The sugar, however, is removed by treatment with acids[51]. Water-soluble compounds may thus strongly interact with rigid and hydrophobic membrane surfaces.

Another method of producing pores on gold surfaces is to apply sulfhydryl bolaamphiphiles **12** with an ω-tetraalkylammonium head group[52]. At short distances, charge repulsion between the head group cancels the Au—S binding forces and the membrane is built with separated alkyl chains (Figure 6.14). Ferricyanide anions can pass through these pores, but not ruthenium cations. The sub-nanometre pores can again be closed by electroneutral amphiphiles.

Finally, one may bind macrocycles to polymeric support material with high permeability. 2 nm pores were prepared in the form of the thiolated calixarene **13** (Figure 6.15)[53,54]. A 12-layer composite of **13** on a gas-permeable polymer[55] separated the He/N_2 gas pair with a permselectivity of 17. The filtration ability of 2 nm pores has also been modelled by a Monte Carlo method[56].

[51] J.-H. Fuhrhop, T. Bedurke, K. Doblhofer, to be published
[52] K. Doblhofer, J.Figura, J.-H. Fuhrhop, *Langmuir*, **1992**, *8*, 1811
[53] M.A. Markowitz, V. Janout, D.G. Castner, S.L. Regen, *J. Am. Chem. Soc.*, **1989**, *111*, 8192
[54] a) M.D. Conner, V. Janout, S.L. Regen, *J. Am. Chem. Soc.*, **1993**, *115*, 1178
b) M.D. Conner, V. Janout, I. Kudelka, P. Dedek, J. Zhu, S.L. Regen, *Langmuir*, **1993**, *9*, 2389
[55] T. Masuda, E. Isobe, T. Higashimura, K. Takada, *J. Am. Chem. Soc.*, **1983**, *105*, 7473
[56] B. Jamnik, V. Vlachy, *J. Am. Chem. Soc.*, **1993**, *115*, 660

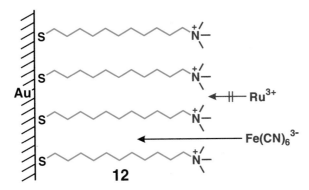

Figure 6.14 *The repulsive interaction between cationic head groups enforces large gaps between sulfur-bound monolayers on gold. Anions can pass such a membrane; cations cannot*[52].

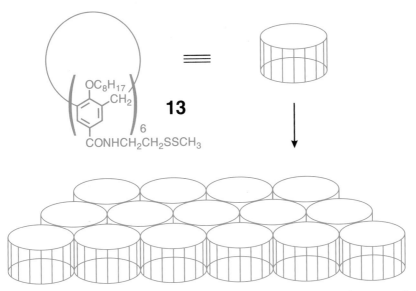

Figure 6.15 *Multilayers of the rigid calixarene macrocycle on porous polymers produce sieves for molecules greater than 2 nm and also serve as membranes for gas separations*[53–55].

A final method of introducing nanopores involves the removal or replacement of "place-holders"[57]. The unsymmetrical disulfide $HO(CH_2)_{16}SS(CH_2)_3CF_3$ reacts with a gold surface to deposit equal amounts of $-S(CH_2)_3CF_3$ and $-S(CH_2)_{16}OH$ groups. The short chain with three methylene groups is replaced by the longer chain $HS-(CH_2)_{16}CN$ on exposure to an ethanol solution of the latter. The replacement of the C_{16} chain takes 10^3 times

---

[57] H.A. Biebuyck, G.M. Whitesides, *Langmuir*, **1993**, *9*, 1766

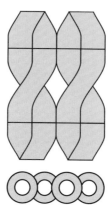

Figure 6.16 *Organization of fully interdigitated, side by side α-helices on the water surface at the inflection point (12.8 Å²/residue for polyalanine and 17.0 Å² for polyleucine). The views are perpendicular and parallel to the water surface.*

longer. Short chain thiols can thus be sequentially removed in order to introduce new guest molecules.

6.5 Polymers, Cast Films and Multilayers

Hydrophobic homopolypeptides are stable in their α-helical conformation at the air/water surface. α-Helices are adsorbed with their long axis parallel to the water surface, the long axis being predominantly perpendicular to the direction of compression. Poly-L-alanine and poly-L-leucine pack together in the form of α-helices on the water surfaces. Inflection points appeared at 12.8 Å² and 17.0 Å² per residue. The α-helices are organized in such a way that their side-chains are interdigitated (Figure 6.16)[58,59].

Polymers can also be formed on the water/air interface by polymerization of water-insoluble hydrophobic monomers. Decoupling hydrophilic spacers which allow partial polymerization of vesicles (see Figure 4.28) are again useful. The monomeric, methacrylic diester **14** spread on water to give a solid phase monolayer (Figure 6.17) up to surface pressures of 50 mN/m. The phase hardly changed upon polymerization (P_M), but polymerization of the same amphiphile in solution gave a polymer which was unable to unfold regularly on the water surface. Only a liquid analogue surface monolayer (P_S) was obtained upon spreading; the lack of a hydrophilic spacer prevented ordering. On the other hand, the same experiment with solution-polymerized amphiphiles e.g. the polymethacrylate made of **15** containing such a spacer produced perfect, solid-like monolayers[60] (not shown).

[58] P. Lavigne, P. Tancrède, F. Lamarche, J.-J. Max, *Langmuir*, **1992**, *8*, 1988
[59] B.R. Malcolm, *Progr. Surf. Membr. Sci.*, **1973**, *7*, 183
[60] A. Laschewsky, H. Ringsdorf, G. Schmidt, J. Schneider, *J. Am. Chem. Soc.*, **1987**, *109*, 7188

Molecular Recognition and Nanopores in Surface Monolayers

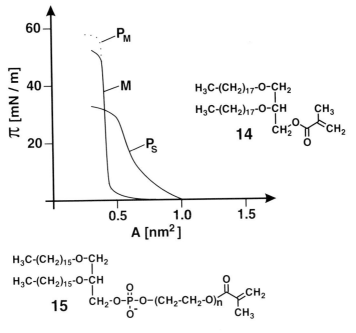

Figure 6.17 *Pressure–molecular area isotherm of monomer* **14** *before* (—) *and after* (···) *polymerization. If the same polymer is formed in solution and spreads, the monolayer is much less well organized (isotherm P_s).*

The polymeric monolayers are as dense as monolayers produced from small molecules. There is a possibility, however, of producing isolated polymer "islands". Monolayers of a polystyrene/poly-4-vinylpyridine AB block polymer, $(PS)_{260}(4\text{-}PVP)_{240}$, fully quaternized with decyl iodide showed regular surface micelles on water with an aggregation number of ~ 120 by transmission electron microscopy (TEM)[61] (Figure 6.18). The poorly water-soluble PVP-part is probably slightly submerged in the subphase but stays at the surface because of its decyl-comb side groups. Essentially 2-dimensional micelles were thus formed (compare with hairy rods in Figure 6.22).

The transfer of polymeric monolayers to substrates is generally not advisable, as polymeric films tend to be too viscous and their transfer rates slow; self-assembly procedures are more promising. Polymerizable multilayers on a solid surface have, for example, been obtained from the dicationic bolaamphiphile **16** and the anionic polyelectrolyte poly(2-acrylamido-2-methyl-1-propanesulfonic acid) **17**. Eight alternating layers of **16** and **17** were self-assembled and stabilized by photopolymerization using UV light. It has been said that organic multilayer films of this type show technological promise in fields such as nonlinear optical devices, sensors and protective coatings[62] (Figure 6.19).

[61] J. Zhu, A. Eisenberg, R.B. Lennox, *J. Am. Chem. Soc.*, **1991**, *113*, 5583
[62] G. Mao, Y. Tsao, M. Tirrell, H.T. Davis, V. Hessel, H. Ringsdorf, *Langmuir*, **1993**, *9*, 3461

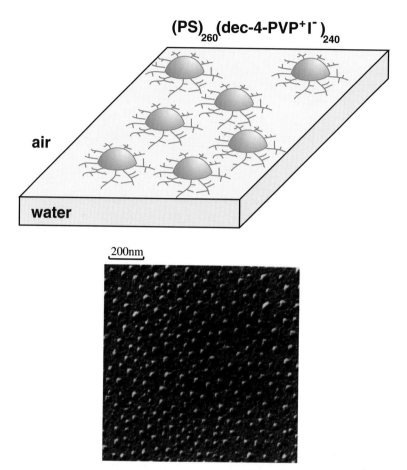

Figure 6.18 *Transmission electron micrograph and model of an LB film of the block polymer between polystyrene and polydecylated polyvinylpyridine isolated at 15 mN/m surface pressure. The hydrophobic blocks are presumbly polystyrene, the amphiphilic spacer arms N-decylpyridinium polymers.*

The molecular ordering of LB films and SAMs can also be transformed to macroscopic ordering by casting aqueous bilayer dispersions (vesicles, myelin figures) onto solid supports. After evaporation of the solid and thermal treatment over several days, cast multilayer films retained their structural characteristics essentially analogous to multilayered LB films. The bilayers were aligned parallelly to the substrate surface and could be removed from the solid support. Free-standing molecular multilayers (= cast films) were obtained. ESR spectroscopy of copper(II) porphyrins revealed that porphyrins with four symmetrically distributed charges can be perfectly aligned parallelly to the membrane surface, whereas membrane-dissolved protoporphyrin IX assumes a more random position (Figure 6.20)[63].

[63] Y. Ishikawa, T. Kunitake, *J. Am. Chem. Soc.*, **1991**, *113*, 621

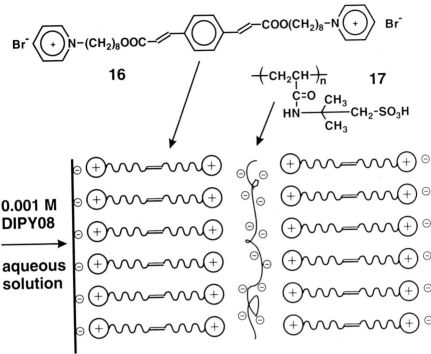

Figure 6.19 *Stepwise synkinesis of multilayer films from polymerizable, dicationic bolaamphiphiles and anionic polyelectrolytes[62].*

Similar self-supporting cast films have also been reported for fullerenes[64], alkyl substituted melanin and isocyanuric acid derivatives[65] and PBr_4^{2-} clusters on continuous bilayers of dodecylammonium counterions[66]. Robust, second-harmonic generation (SHG; frequency doubling of laser light) materials were produced by an orienting and protecting hyperpolarizable amphiphilic dye between solid layers of hydrophobic (polystyrene) and hydrophilic (poly-ethylene—CO—maleic acid) polymers. The pyridinium charge stuck to the succinic acid and allowed rinsing with toluene to remove excess dye. The alkyl chain then dissolved in a coating of polystyrene (Figure 6.21). A sandwiched-monolayer formed showing strong SHG activity[67].

The synkinesis of isolable organized monolayers by the LB technique is not restricted to amphiphiles. They can also be obtained from stiff polymers with multiple hydrocarbon side-chains ("hairy rods")[68a]. Regular and stable bilayers were obtained consisting of stiff rods made of cellulose or silicon phthalocyanate chains which were separated by fluid alkane regions. Furthermore the

[64] C. Jehoulet, Y.S. Obeng, Y.-T. Kim, F. Zhau, A.J. Bard, *J. Am. Chem. Soc.*, **1992**, *114*, 4237
[65] N. Kimizuka, T. Kawasaki, T. Kunitake, *J. Am. Chem. Soc.*, **1993**, *115*, 4387
[66] N. Kimizuka, T. Maeda, I. Ichonose, T. Kunitake, *Chem. Lett.*, **1993**, 941
[67] J.P. Gao, G.D. Darling, *J. Am. Chem. Soc.*, **1992**, *114*, 3997
[68a] M. Schaub, G. Wenz, G. Wegner, A. Stein, D. Klemm, *Adv. Mater.*, **1993**, *5*, 919

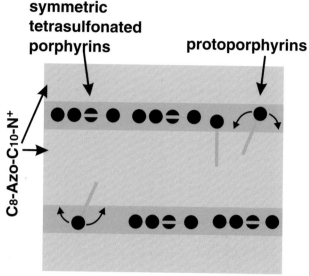

Figure 6.20 *Lipid multilayers in cast films may be doped with water-soluble or amphiphilic porphyrins. The hydrophobic macrocycle of protoporphyrin derivatives (—) with one hydrophilic edge integrates into the hydrophobic membrane, but their orientation is not fixed. Water-soluble porphyrins with four sulfonated phenyl rings at the methine bridges are oriented parallel to the film surface. The ordering of the hydrophobic bilayer can also be controlled by azo-dyes which are covalently bound to the hydrophobic chains of the lipids, here C_8H_{17}—C_6H_4—$N\!=\!N$—C_6H_4—$C_{10}H_{20}$—$N(CH_3)_3Br$.*

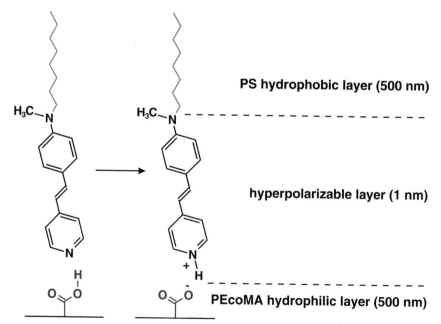

Figure 6.21 *Orientation of a hyperpolarizable dye layer at the surface of a hydrophilic copolymer[67].*

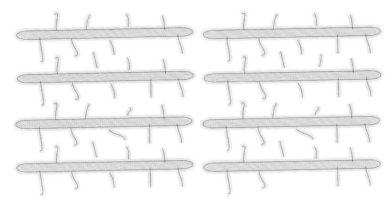

Figure 6.22 *Schematic picture of "hairy rods" as applied on solid surfaces by the Langmuir–Blodgett (LB) technique. The hairs are alkyl chains, not amphiphiles. The rods are cellulose, polypeptides or other stiff polymers. Typical surface pressures are around* 20–30 mN/m.

rods are strictly oriented in a direction perpendicular to the water surface. If the direction of the LB plate is not changed, all the rods will be parallel (Figure 6.22); if it is turned by 90° in each application, a grid will be formed.

6.6 Bolaamphiphiles and Dyes on Solid Surfaces

Boolaamphiphiles with two reactive head groups at both ends may bind in a linear or bent fashion to metal surfaces. The only example of a folded bolaamphiphiles on solid surfaces is provided by $HOOC-(CH_2)_{30}-COOH$ on silver. Ellipsometry yielded a surface film thickness of 20 ± 2 Å. In the reflection infrared spectra, the 2928 cm^{-1} shoulder, as well as the lack of sharp bands between 1150 and 1350 cm^{-1} for the progression of wagging modes indicated that the *all-trans* sequence length in the surface film should be much less than for the fully extended chain in the KBr pressed disk sample (Figure 6.23)[68b].

In all other cases, a linear conformation was found; a fact also be seen for α,ω-dithiols[69-71]. The C_{16} compound, $HS-(CH_2)_{16}-SH$, can be employed in the binding of metal ions to the surface of gold electrodes[69]. 1,6-Hexanedithiol was used to fixate cadmium sulfide nanocrystals on the gold surface[70] (Figure 6.24). More rigid staffane analogues fixate Ru^{2+} complexes and the properties of the electroactive films depend on the pH[71] (no Figure).

One or two additional functional groups introduced via other α-hydrosulfide bolaamphiphiles can be characterized by measurements of the contact angle θ_a of water and hexadecane (= HD) droplets, depending on the hydrophobicity and the position of the end groups (Table 1). They also attest to the short-range

[68b] D.L. Allara, S.V. Atre, C.A. Elliger, R.G. Snyder, *J. Am. Chem. Soc.*, **1991**, *113*, 185
[69] J. Figura, Ph.D Thesis, Freie Universität Berlin, **1992**
[70] V.L. Colvin, A.N. Goldstein, A.P. Alivisatos, *J. Am. Chem. Soc.*, **1992**, *114*, 5221
[71] Y.S. Obeng, M.E. Laing, A.C. Friedli, H.C. Yang, D. Wang, E.W. Thulstrup, A.J. Bard, J. Michl, *J. Am. Chem. Soc.*, **1992**, *114*, 9943

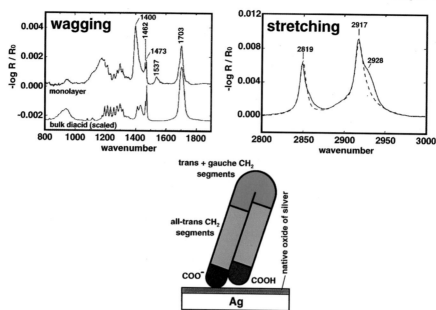

Figure 6.23 *Infrared spectra and a model of a HOOC—(CH$_2$)$_{30}$—COOH monolayer on silver. A comparison with spectra of* all-trans *configured bulk acid (lower and dotted traces) clearly indicate the* gauche *bend in the monolayer (compare with Figure 5.16).*

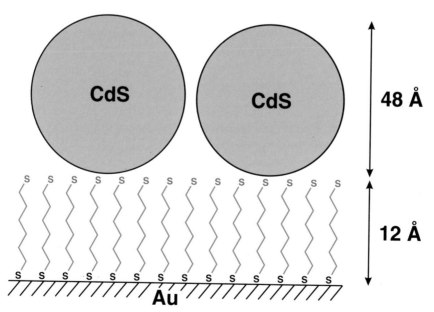

Figure 6.24 *Long chain α,ω-dithiols bind with only one SH group to gold. The remaining outer SH group can then be used to fixate metal colloids, transition metal ions or electrophilic organic substrates.*

Table 1 Advancing contact angles θ_a on thiol monolayers adsorbed on gold.

RSH	θ_a H_2O	HD	RSH	θ_a H_2O	HD
$HS(CH_2)_2(CF_2)_5CF_3$	118	71	$HS(CH_2)_{11}OCH_3$	74	35
$HS(CH_2)_{21}CH_3$	112	47	$HS(CH_2)_{12}SCOCH_3$	70	0
$HS(CH_2)_{17}CH\!=\!CH_2$	107	39	$HS(CH_2)_{10}CO_2CH_3$	67	28
$HS(CH_2)_{11}OSi(CH_3)_2C(CH_3)_3$	104	30	$HS(CH_2)_8CN$	64	0
$HS(CH_2)_{11}Br$	83	0	$HS(CH_2)_{11}OH$	0	0
$HS(CH_2)_{11}Cl$	83	0	$HS(CH_2)_{15}CO_2H$	0	0

nature of the molecular interactions responsible for wetting[72]. Mixed monolayers with partly buried hydroxyl groups switched from hydrophobic to hydrophilic[73] at a hydrocarbon/alcohol range of 1:10; terminal ether linkages are solvated by water down to 5 Å beneath the surface[74].

ω-Terminated alkanethiols with CN, OH and COOH groups also form oriented monolayers on copper, silver and gold surfaces. The cant angles of the *trans*-extended chains are 28° on gold and 13° on copper and silver[75]. Mixed monolayers of 1-hexadecanethiol and 16-thiohexadecan-1-ol showed no detectable hydrogen bonding between the OH groups, indicating molecular mixing. *In situ* derivatization with trifluoroacetic anhydride in the presence of triethylamine produced quantitative yields of the trifluoroacetate[76]. Phosphates and sulfates have also been described. Long-lived bilayers were prepared by the reaction of SAMs containing terminal carboxylic groups with n-alkylamines in the vapour phase. Surprisingly enough, a significant portion of protonated COOH groups survived the attack by the amine (Figure 6.25), probably a result of the repulsive Coulombic force between neighbours in the same plane carrying the same charge. Atoms in different planes are normally further away[77].

A series of dyes were dissolved in fluid surface monolayers or were covalently attached to the surface; several typical examples follow and other examples and possible applications are discussed in Ulman's book[1].

The fluidity of LB monolayers on solid supports was characterized by measuring the long-range (> 10 μm) translational diffusion of dissolved fluorescent lipids using fluorescence pattern photobleaching recovery (FPPR). In this technique, an area of the membrane is illuminated with a spatially striped intensity. After irreversibly photobleaching, fluorescence recovery occurs, as unbleached molecules from nonilluminated stripes move into the illuminated stripes[78]. In fluid-like LB films, translational diffusion coefficients

[72] C.D. Bain, E.B. Troughton, Y.-T. Tao, J. Evall, G.M. Whitesides, R.G. Nuzzo, *J. Am. Chem. Soc.*, **1989**, *111*, 321
[73] C.D. Bain, G.M. Whitesides, *J. Am. Chem. Soc.*, **1988**, *110*, 3665
[74] C.D. Bain, G.M. Whitesides, *J. Am. Chem. Soc.*, **1988**, *110*, 5897
[75] P.E. Laibinis, G.M. Whitesides, *J. Am. Chem. Soc.*, **1992**, *114*, 1990
[76] L. Bertilsson, B. Liedberg, *Langmuir*, **1993**, *9*, 141
[77] L. Sun, R.M. Crooks, A.J. Ricco, *Langmuir*, **1993**, *9*, 1175
[78] B.A. Smith, H.M. McConnell, *Proc. Natl. Acad. Sci. U.S.A*, **1978**, *75*, 2759

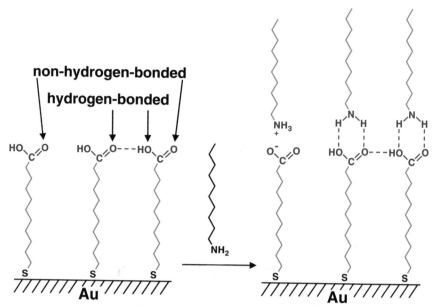

Figure 6.25 *Surface amino groups adsorb onto both protonated and deprotonated carboxylic acids.*

are $\sim 10^{-8}$ cm^2/s; in solid-like films they drop to $\leq 10^{-10}$ cm^2/s. Similar techniques can also be applied to differentiate between fluorescently labelled proteins which are in equilibrium between solution and binding sites on the LB layer. Dynamics of antibody binding to antigenes and to the monolayer have thus been elucidated in detail[79] using laser-based, time-resolved fluorescence microscopy.

Cysteamine was used to couple redox-active carboxylalkyl-4,4'-bipyridinium salts to the gold surface[80]. The nonordered monolayer assembly was then transformed into a densely packed monolayer with 1-hexadecanethiol and cyclic voltammetry of the surface bound viologen was performed. The electron transfer rate constants to the bipyridinium sites depended on the alkyl chain length d bridging the redox site to the electrode. Electron transfer rate constants k_t followed the Marcus theory[81]. Cysteic-acid-active ester monolayers chemisorbed on gold were used to electrode-immobilize the protein glutathione reductase, then a bipyridinium carboxylic acid was condensed onto the enzyme in the presence of urea to "wire" the protein towards electrochemical reduction[82] (Figure 6.26).

7,7,8,8-Tetracyanoquinodimethane, tetrathiafulvalene, *p*-phenylenediamine

---

[79] N.L. Thompson, C.L. Poglitsch, M.M. Timbs, M.L. Psarchick, *Acc. Chem. Res.*, **1993**, *26*, 567
[80] E. Katz, N. Itzhak, I. Willner, *Langmuir*, **1993**, *9*, 1392
[81] R.A. Marcus, N. Sutin, *Biochim. Biophys. Acta*, **1985**, *811*, 265
[82] I. Willner, E. Katz, A. Riklin, *J. Am. Chem. Soc.*, **1992**, *114*, 10963

Molecular Recognition and Nanopores in Surface Monolayers 179

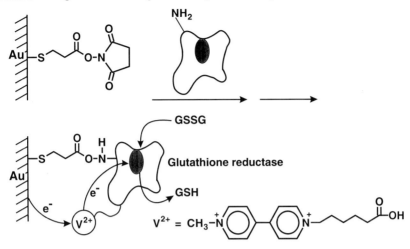

Figure 6.26 *Instead of ATP and a biological redox chain one can also use an electrode and a chemical viologen "wire" to energetize an enzymatic reaction, e.g. reduction of the glutathione dimer.*

and anthraquinone were employed as head groups for amphiphiles with oligomethylene chains or cholesterol as the hydrophobic part. In combination with a large hydrophilic group, the steroid skeleton was most effective in the formation of non-viscous films and uniform LB surface monolayers[83]. Dithiol functionalized anthraquinone was effectively bound to gold and in the presence of a 1-alkanethiol (C_{10}–C_{18}) gave reversible peaks in cyclic voltammetry[84] (Figure 6.27).

Porphyrin monolayers at the air/water interface were handled extensively in Ulman's book[1]. A self-assembly technique for depositing porphyrin monolayers on quartz and silicon surfaces has recently been developed. (Chloromethyl)phenyltrichlorosilane was first used for obtaining benzyl chloride coated surfaces which were then coupled with *meso*-tetrapyridylporphyrin (Figure 6.27). Washing and quaternization with iodomethane yielded cationic porphyrin surfaces with 1 porphyrin per 1.04 nm^3. FTIR–ATR spectroscopy with both *p*- and *s*-polarized incident polarization, as well as nonlinear optical properties using second-harmonic generation, indicated an average polar angle of 43° and a monolayer thickness of 17 Å [85].

Finally, the unavoidable fullerenes were, of course, immobilized on solid surfaces. Fullerene, C_{60}, and fullerene epoxide formed stable monolayers on water. The limiting molecular surface area was 95 Å and a UV/VIS spectrum with bands around 300 nm was observed in LB multilayers on quartz[86,87].

[83] K. Naito, A. Miura, M. Azuma, *J. Am. Chem. Soc.*, **1991**, *113*, 6386
[84] L. Zhang, T. Lu, G.W. Gokel, A.E. Kaifer, *Langmuir*, **1993**, *9*, 786
[85] D.Q. Li, B.I. Swanson, J.M. Robinson, M.A. Hoffbauer, *J. Am. Chem. Soc.*, **1993**, *115*, 6975
[86] Y.S. Obeng, A.J. Bard, *J. Am. Chem. Soc.*, **1991**, *113*, 6279
[87] N.C. Maliszewskyj, P.A. Heiney, D.R. Jones, R.M. Strongin, M.A. Cichy, A.B. Smith III, *Langmuir*, **1993**, *9*, 1439

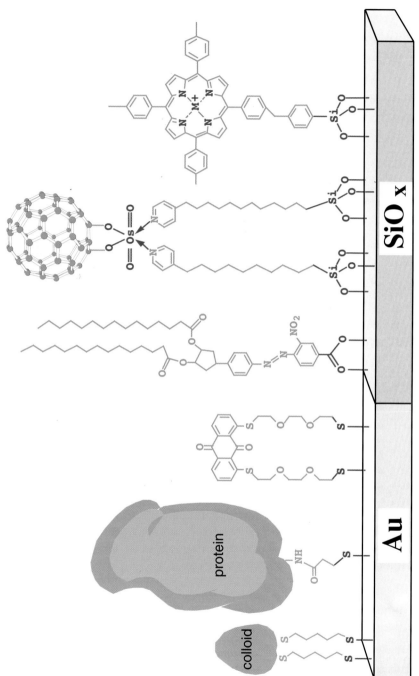

Figure 6.27 *Various reactive head groups which have been covalently bound on the surface of SAMs. See text and references.*

Deposition of fullerene "monolayers" on glass by a subphase-lowering method and subsequent replication in polymer films gave TEM images of C_{60} clusters with an average diameter of 52.2 Å and a thickness corresponding to six linearly chained C_{60} molecules[88]. Fullerene osmate was covalently attached to self-assembled, pyridyl-terminated monolayers on quartz or Ge/Si multilayer substrates (Figure 6.27)[89]. Similar fullerene monolayers were also obtained by a spontaneous assembly of fullerene on gold-fixated cysteamine monolayers[90].

[88] M. Iwahashi, K. Kikuchi, Y. Alchiba, I. Ukemoto, T. Araki, T. Mochida, S. Yokoi, A. Tanaka, K. Iriyama, *Langmuir*, **1992**, *8*, 2980
[89] J.A. Chupa, S. Xu, R.F. Fischetti, R.M. Strongin, J.P. McCauley, A.B. Smith III, J.K. Blasie, L.J. Peticolas, J.C. Bean, *J. Am. Chem. Soc.*, **1993**, *115*, 4383
[90] W.B. Caldwell, K. Chen, C.A. Mirkin, S.J. Babinec, *Langmuir*, **1993**, *9*, 1945

CHAPTER 7

Amphiphilic Crystals and Hydrogen Bonded Co-Crystals

7.1 Introduction

So far, we have described synkinesis only for molecular mono- and bilayer systems. 3D crystals are, in general, considered to constitute stable, chemically dead systems with no potential for the construction of reactive, supramolecular systems. There are, however, exceptions. First of all, the surface of crystals is, of course, as reactive as any other surface. Crystal engineering is considered here as the "solid-state branch" of synkinesis. Furthermore crystals with large cavities have recently been prepared. They contain "inner surfaces" which may have interesting receptor properties in co-crystallization and photochemical fixation processes, which constitute another type of planned synkinesis. Furthermore, spontaneous 3D crystallization may compete with the synkinesis of membranes and the molecular conformations and interactions in crystals are important standards for the study of membranes. The study of 3D crystal structures of amphiphiles is therefore mandatory as a basis for all structural work on molecular assemblies.

7.2 Chiral and Crystalline Langmuir Monolayers on Water

As a first approximation, molecules in surface monolayers can be considered as being freely rotating alkane chains. With cylindrical symmetry, these cylinders may pack together in three different cells: hexagonal, centred rectangular (= orthorhombic) and oblique (= triclinic). In a hexagonal cell, all alkane chains are aligned vertically relative to the plane of the monolayer; in the other cells they are tilted. All Langmuir films studied to date are not single crystals, but powders of randomly ordered 2D crystallites. Nevertheless one often measures experimentally **high crystallinity in the monolayer's uncompressed state** and a tendency towards a reduction of crystallinity with increased pressure. GID (= grazing incidence diffraction) measurements on phospholipid monolayers, for example, yielded a single peak in the pressure range between 10 and 38 mN/m, but at high surface pressures, in the so-called solid-like state, the

Amphiphilic Crystals and Hydrogen Bonded Co-Crystals

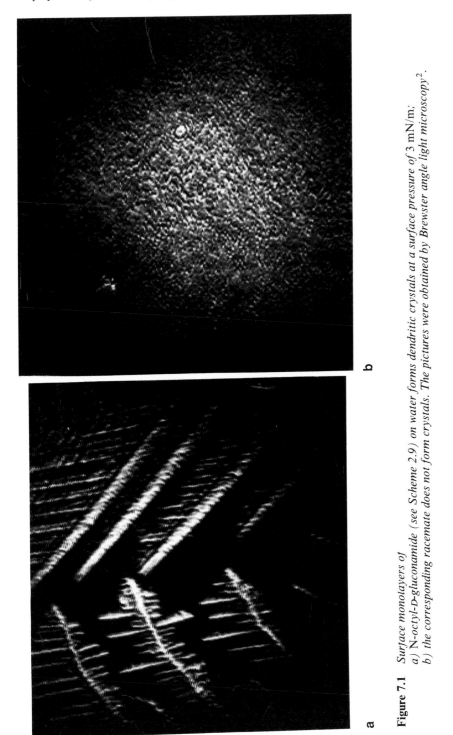

Figure 7.1 *Surface monolayers of*
a) N-octyl-D-gluconamide (see Scheme 2.9) on water forms dendritic crystals at a surface pressure of 3 mN/m;
b) the corresponding racemate does not form crystals. The pictures were obtained by Brewster angle light microscopy[2].

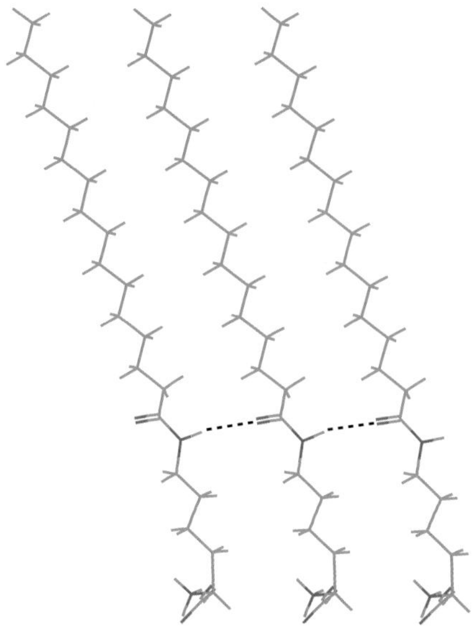

Figure 7.2 *Palmitoyl-R-lysine monolayers on water are not only stabilized by binding interactions between the hydrocarbon chains and amino acid head groups, but linear amide hydrogen bonds between the ε-amide groups are also formed[3]. This requires a 30° tilt.*

coherence length was no more than 50 lattice distances[1]. *N*-Octyl-D-gluconamide forms dendritic monolayer crystals within a few seconds on water at a surface pressure of about 5 mN/m. Below this pressure, the crystals slowly dissipate. Above 30 mN/m, the crystals merge into each other and break down to form a microcrystalline powder[2]. The racemic mixture of the D- and L-gluconamide gave no dendrites (Figure 7.1a,b; compare with Figures 6.1 and 6.2). The homodromic hydrogen bond cycle found in 3D crystals (Figure 7.12) presumably also exists in the monolayers.

ε-Palmitoyl-*R*-lysine monolayers are hydrogen bonded by the α-amino acid head group NH_3^+—CH—COO^- as well as by the amide groups CO—NH in the chain (Figure 7.2). At 25 mN/m, domains with a coherence length of 50 nm were found. Only a small amount of the opposite enantiomer causes the GID peak to disappear; as little as 3% leads to serious misalignment of the 2D crystal sheet, the pure racemate gave no diffraction pattern. The corresponding 3D crystals of *rac*-α-amino acids often showed arrangements in which *R* and *S* molecules were linked in a layer by glide symmetry[3]. This arrangement is obviously unstable in monolayers.

On the other hand, perfluorododecylaspartate monolayers have a relatively

---

[1] D. Jacquemain, G. Wolf, F. Leveiller, M. Deutsch, K. Kajaer, J. Als-Nielsen, M. Lahav, L. Leiserowitz, *Angew. Chem.*, **1992**, *104*, 134; *Angew. Chem., Int. Ed. Engl.*, **1992**, *31*, 130
[2] K. Vollhardt, J.-H. Fuhrhop, to be published
[3] I. Weissbuch, L. Addadi, L. Leiserowitz, M. Lahav, *J. Am. Chem. Soc.*, **1988**, *110*, 561 (see also *J. Am. Chem. Soc.*, **1990**, *112*, 7718)

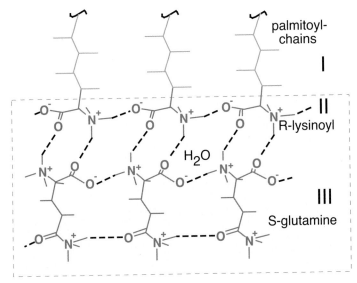

Figure 7.3 *Reflectivity data of the palmitoyl-R-lysine monolayer (see Figure 7.2) in the presence of S-glutamine in the subphase show that about one quarter of the sites below the lysine head groups are occupied by glutamine. Amide hydrogen bonds cause the enrichment of glutamine on the surface; charge interactions between the α-amino acid moieties also stabilize the hydrophilic glutamine layer*[1].

large molecular area of 28.5 Å^2 (in comparison to 24.3 Å^2 for palmitoyllysine with CH$_2$ chains) and their hydrogen bond chains are more loosely connected. The open head group arrangement tends to abide with the observation that racemic mixtures display an intense GID diffraction peak. Langmuir monolayers of fluorinated molecules are often characterized by a higher crystallinity than their hydrocarbon counterparts because the large fluorine atoms render the chains stiffer[1].

Reflectivity measurements of palmitoyl-D-lysine monolayers over dilute solutions of L-glutamine showed that as many as a quarter of the sites below the lysine head groups are occupied by glutamine. Intermolecular hydrogen bonding is thus very strong at the air/water interface (Figure 7.3), which was also established by carbohydrate binding studies (see section 6.2).

7.3 Solubilities and Single Crystals of Achiral Amphiphiles

Single 3D crystals alone are of limited interest in supramolecular chemistry since their surface is small and the large majority of their monomers is inaccessible to chemical reactions with external molecules. Nevertheless, X-ray diffraction on single crystals is a major tool of supramolecular assembly chemistry, possessing the most detailed information on molecular interactions from crystal structures. If one compares solid state infrared and NMR spectra of curved supramolecular systems with those of 3D crystals, one can deduce molecular conformations in the microcrystalline assemblies (see section 5.5). **A crystal structure defines the elementary binding properties of a synkinon under tightly packed conditions.**

The simplest and most common synkinons are non-branched, saturated fatty acids from C12 to C18 (trivial names: lauroyl C12, myristoyl C14, palmitoyl or cetyl C16, stearoyl C18) and their sodium, ammonium and potassium salts (also known as "soaps"). Lauric, myristic, palmitic and stearic acids are barely soluble in water at 20°C (5.5, 2.0, 0.7 and 0.3 mg/L) and 60°C (8.7, 3.4, 1.2 and 0.5 mg/L), each ethylene group lowering the solubility by a factor of 2–3. The solubilities of the corresponding sodium and potassium salts are, however, in the order of several grams per litre. Even in highly concentrated emulsions of soaps in distilled water ($\geq 30\%$ w/w), precipitation of solids is often not observed. Bivalent fatty acid salts, however, are just as insoluble as free fatty acids: only 1.4 mg of calcium stearate dissolves in 1 L of water[4].

In their *all-trans* conformation, fatty acids and soaps possess the typical shape of a good synkinon (see figure 1.1), i.e. they are long, thin and assemble well along the oligomethylene chains via additive van der Waals interactions, but are found to interact negligibly at the methyl endings within the bilayer. **The weakness of the hydrophobic end-to-end interactions slows down the formation of regular 3D crystals dramatically, whereas isolated bilayers (= 2D crystals) with frequent irregularities rapidly build up through the hydrophobic effect.**

There is, however, no certainty as to whether intralayer or interlayer interactions between the head groups enforce crystalline order between these bilayers and whether they also determine supramolecular structures or not. These interactions must be established experimentally in each individual case. The carboxyl groups, for example, form strong, bifurcated hydrogen bonds and enforce interactions between the crystal sheets (Figure 7.4a)[5]. On the other hand, charged carboxylate groups of soaps reject each other, causing the soap crystal sheets to be extremely slippery. So far, only one single long-chain carboxylate with a monovalent counterion (potassium palmitate) has been crystallized[6]. The potassium ions lie between the carboxylate groups in a crystal sheet in the centre of the facing carboxylate groups (Figure 7.4b).

Bivalent metal salts are normally too insoluble to be crystallized, but here, a reasonable compromise has been seen in organic bis-ammonium counterions. Piperazinium counterions of fatty acid salts connect two crystal sheets via a rigid cyclohexane unit producing large, well-formed crystals[7]. The four NH groups of each cation also form hydrogen bonds with a carboxylate oxygen of different alkanoate anions and thus form a cationic layer between the anionic sheets. An unexpected result of the arrangement of these counterions

---

[4] D.M. Small, The Physical Chemistry of Lipids, Plenum Press, New York, **1986**
[5] F. Kaneko, M. Kobayashi, Y. Kitagawa, Y. Matsuura, *Acta Cryst.*, **1990**, *C46*, 1490
[6] J.H. Dumbleton, T.R. Lomer, *Acta Cryst.*, **1965**, *19*, 301
[7] F. Brisse, J.-P. Langin, *Acta Cryst.*, **1982**, *B38*, 215

Amphiphilic Crystals and Hydrogen Bonded Co-Crystals

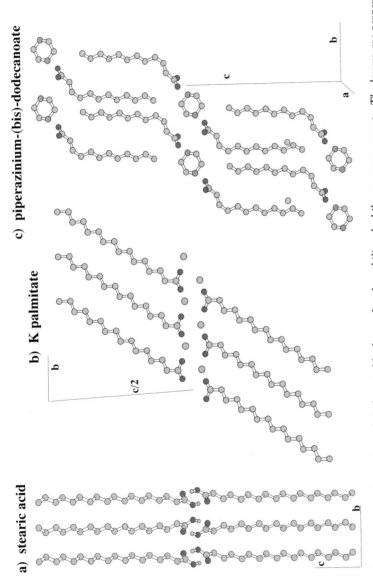

Figure 7.4 Crystal structures of a) stearic acid. Bifurcated hydrogen bonds stabilize the bilayer arrangements. The layers are perpendicular to the carboxyl plane in this particular crystal form, although the COOH groups are larger than CH_2 groups[5]. b) potassium palmitate. Soaps are very difficult to crystallize, because there is no binding interaction between the crystal sheets[6]. c) piperazinium bis-dodecanoate. The crystallization becomes more favoured if a fitting dication connects the carboxylate chains[7]. The chains interdigitate, which is probably triggered (i) by the cyclic counterion, which also consists of two parallel alkyl moieties, and (ii) by the repulsion between the symmetrically charged carboxylate groups.

between the crystal sheets is the replacement of the tail-to-tail bilayer (Figures 7.4a,b) by an interdigitated bilayer (Figure 7.4c).

Views of the ab-plane onto the head group and methyl group regions of the same structures show the cross-sectional areas of both the oligomethylene chains and the carboxyl end groups including the counterions. The head group cross section area is usually somewhat wider than the oligomethylene area and is balanced by a tilt of the hydrocarbon chains towards the carboxylate chains and/or by interdigitating of the chains. A comparison of the three structures shows a similar inclination of the alkyl chains towards the carboxylate plane (60–70°) and no significant deviation from the *all-trans* conformation. The only important difference is the interdigitation which cuts the bilayer thickness to about 60% of the tail-to-tail bilayer.

Another interesting detail concerns **the lengths of the carbon oxygen bonds**. They **are 1.14 Å for the single bond and 1.39 Å for the double bond in the case of stearoyl ethyl esters** (see next paragraph). The lengths of the same double and single bonds are **1.21 and 1.27 Å in free stearic acid and palmitic potassium salts**. There was no significant difference in bond lengths to be found **in the symmetric, hydrogen bonded piperazinium palmitate where the bond lengths are 1.25 and 1.26 Å**. One may therefore expect the **strongest carboxylate repulsion** within one crystal sheet to take place **in the piperazinium salt** where both oxygen atoms are negatively charged, thus **opening the space required for interdigitation**.

Ethyl stearate (Figure 7.5) and higher esters form single molecular layer crystal sheets with a regular alignment of all molecules in the same direction (head-to-tail)[8]. One particular case is that of methyl esters. Although there is no possibility of hydrogen bonding between the head groups of the crystal sheets of *methyl* stearate[9], the molecules arrange in the same way, in tail-to-tail bilayers. The closest distance between the ester carbons becomes as short as 3.3 Å, indicating strong polar forces between carboxyl oxygen and carbon atoms. These forces are probably the cause of dimerization and have the same effect as hydrogen bonding.

Saturated fatty acids exhibit polymorphism based mainly on the various hydrocarbon chain packings depicted in Figure 7.6. Various thermal rearrangements are therefore common in the melting process, but the accompanying occurrence of differing molecular conformations is more significant. For example, stearic acid appears with an *all-trans* conformation causing the α-carboxyl and methyl groups to be located in the same layer. In the orthorhombic type of crystal, the bond between C(2) and C(3) is *gauche* configured. Carboxyl and methyl groups are then found in separate layers (no Figure). The crystalline states of hydrocarbon chains are conveniently classed as subcells in which a 2D lattice a·b lies perpendicularly to the chain's axis and c is two carbon atoms long (= 2.5 Å). Triclinic, orthorhombic and hexagonal subcells result (Figure 7.6)[4]. In triclinic and orthorhombic packings, the mean volume of the CH_2 groups is 24 Å^3, the surface area 19 Å^2. The hexagonal packing is

---

[8] S. Aleby, *Acta Chem. Scand.*, **1968**, *22*, 811
[9] S. Aleby, E. von Sydow, *Acta Cryst.*, **1960**, *13*, 487

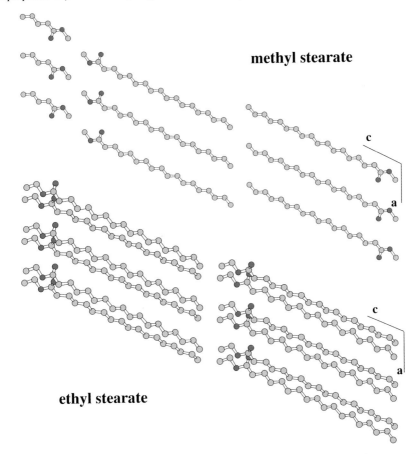

Figure 7.5 *The crystal structure of methyl stearate shows a tail-to-tail orientation, whereas the corresponding ethyl ester is head-to-tail oriented*[8,9].

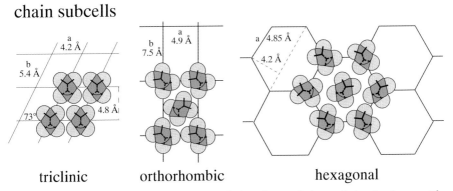

Figure 7.6 *Three frequent arrangements of the oligomethylene chains in fatty acid crystals and derived lipids. The hexagonal packing is less tight than the other two and gives rotational freedom to chains and head groups*[4].

not so tight with 26 Å^3 and 20 Å^2 for each CH$_2$ group. Odd-numbered hydrocarbons undergo an orthorhombic-to-hexagonal transition before melting. The mean volume of CH$_2$ groups in the liquid state rises to 30 Å^3, the mean surface to 23 Å^2. The crystal structures of fatty acid esters, on the other hand, show a simple layer structure.

Certain fatty acids contain *cis*-configured double bonds which cause a distortion of the chain's linearity around the double bond. Very often, two *gauche* conformations of the oligomethylene chain in the neighbourhood of the double bond are used to linearize the chain as in the case of linoleic acid (Figure 7.7a)[10]; in oleic acid crystals, however, a U-shaped conformer was isolated (Figure 7.7b)[11]. The oligomethylene chains are *all-anti* configured. Another important effect of the methine carbon atoms in unsaturated fatty acids in a rise in the flexibility of the neighbouring methylene groups. Rotation becomes much easier and crystallization occurs only at low temperatures.

The cationic counterparts of fatty acids are long chain ammonium salts with single, double or triple alkyl chains. Tetraalkylammonium and *N*-alkylpyridinium amphiphiles possess approximately half the solubilities and cmc's of their anionic carboxylate counterparts. They can be precipitated with perchlorate or iodide counterions which destroy salt solubilizing water clusters (= "chaotropic" effect). Ammonium ions share this property with potassium ions which have equally weak hydration energies. The crystal structure of dodecyltrimethylammonium bromide corresponds to a combination of potassium palmitate (side by side arrangement of ions and counterions) and piperazinium palmitate (interdigitation). The double-chain amphiphile didodecyldimethylammonium bromide shows a normal tail-to-tail arrangement in crystals[12]. The deviation of the alkyl chains from the 90° angle to the ammonium plane is 18° in the single-chain amphiphile and 40° in the double-chain analogue (Figure 7.8a,b). An ammonium salt with a long dodecyl side-chain, a short propyl chain and two methyl groups also forms stable crystals but contains empty spaces (not shown).

Amphiphiles with a tetraalkylammonium head group, a central azobenzene unit and two alkyl chains with n and m methylene groups each produce different crystal structures depending on m and n. When both chains are of approximately equal length ($m = 8$, $n = 10$), interdigitated chains (Figure 7.8c) lying perpendicularly to the plane of the head groups are formed[13]. If, however, the chain between the cationic head group and the rigid segment becomes very short ($m = 5$) and the terminal chain is long ($n = 12$), interdigitation becomes impossible, as the chain segments no longer fit. A double-layer is then formed. X-ray photoelectron spectroscopy (XPS) of such crystals revealed a 30° angle between the chain and head group planes[14]. This tilt adjusts for the bulkiness of

[10] J. Ernst, W.S. Sheldrick, J.-H. Fuhrhop, *Z. Naturforsch.*, **1979**, *34b*, 706
[11] S. Abrahamsson, I. Ryderstedt-Nahringbauer, *Acta Cryst.*, **1962**, *15*, 1261
[12] K. Okuyama, Y. Soboi, N. Iijima, K. Hirabayashi, T. Kunitake, T. Kajiyama, *Bull. Chem. Soc. Jpn.*, **1988**, *61*, 1485
[13] G. Xu, K. Okuyama, M. Shimomura, *Mol. Cryst. Liq. Cryst.*, **1992**, *105*, 213
[14] K. Okuyama, H. Watanabe, M. Shimomura, K. Hirabayashi, T. Kunitake, T. Kajiyama, N. Yasuoka, *Bull. Chem. Soc. Jpn.*, **1986**, *59*, 3351

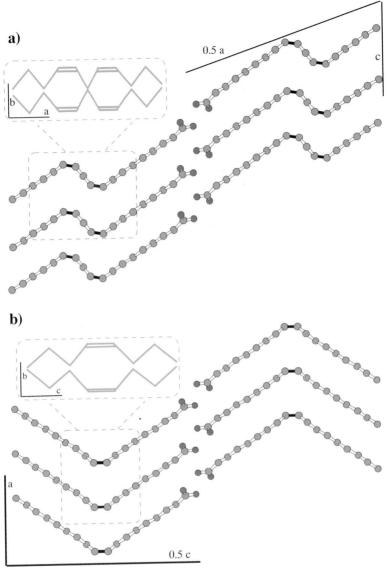

Figure 7.7 *The two double bonds in linoleic acid (a) induce* gauche *orientations in the neighbouring oligomethylene chains. The U-shape of oleic acid (b) is thus avoided*[10,11].

the head groups. The azobenzene units shifted against each other. In the first case, both chromophores faced each other (H-aggregate, Figure 7.8c); in the second case they were also close to each other but in a more lateral position (J-aggregate, Figure 7.8d). The first crystal's long wavelength band was at 300 nm, the second at 390 nm.

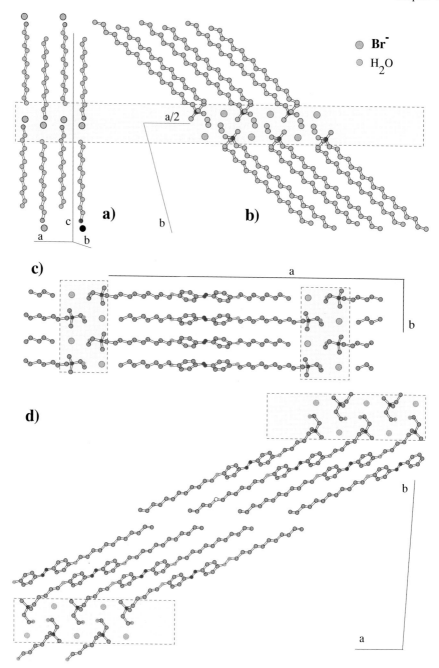

Figure 7.8 *Crystal structures of straight single-chain (a) and tilted double-chain (b) ammonium salts. Similar single-chain ammonium salts with azobenzene units in the centre interdigitate, so that the chains lie perpendicular to the head group area (c); or do not interdigitate, which introduces a 30° angle (d). As a result the chromophores form J- or H-aggregates*[12–14].

7.4 Single Crystals of Chiral Amphiphiles

Chiral head groups of amphiphiles, e.g. carbohydrates, tartaric acid or amino acids, introduce a symmetry problem into the crystallization process. If such an amphiphile crystallizes in the usual bilayer assembly, the chiral head groups meet in inverse orientations. Each substituent at a chiral centre which is up-side oriented in the left-hand molecule, will be down in the right-hand counterpart. This situation could be problematic to attain in a planar crystal sheet. On the other hand, regular ordering of all chiral centres asks for an equally unfavourable head-to-tail arrangement where polar head groups neighbour apolar alkyl chains. Both situations (tail-to-tail and head-to-tail arrangements) were found in crystals of chiral amphiphiles, as well as the intermediate case of "interdigitated" arrangements.

Amphiphiles with chiral head groups are of vital interest to synkinesis as they do not only produce the most stable and complex fibres, but they are also a presupposition for stereoselective recognition processes. Tartaric acid provides two chiral centres with hydroxyl groups in addition to two achiral carboxyl groups. The hydrophobic chain can be connected as a secondary amide to one end to form amphiphilic tartaric acid monoamides. In the crystal sheets of two of these compounds, the amphiphiles are not only connected by van der Waals interactions, but are also directed by hydrogen bonds between the hydroxyl groups. No amide hydrogen bonding was observed. Ionic interactions determine the ordering of the monomers; the sodium ion binds four[15], the ammonium ion two carboxylate groups[16]. A head-to-head arrangement is thus enforced in both the pure enantiomer and the racemic crystals. It was also found that the racemate would only crystallize in the form of a 2-monoacetate. Here, the terminal methyl group of the octyl chain of one enantiomer connects with the methyl group of the acetyl side-chain of its counterpart. The direct fitting together of the hydrophobic chain of one molecule with the hydrophilic head group of its neighbour is thus achieved in an interdigitated and racemic bilayer of two bent molecules (Figure 7.9).

At this point we also have to consider the many crystal structures of non-amphiphilic enantiomers and diasteromers of tartaric acids, their salts, double salts and esters. In general, metal ions are highly hydrated and complicated hydrogen bond chains connect the carboxyl and hydroxyl groups (Figure 7.10a). An interesting speciality can be found in racemic acid without metal counterions. Each hydroxyl group forms two hydrogen bonds to other hydroxyls, one being on a D- and the other one an L-molecule so that a square is formed by the hydroxyl groups of four molecules[17] (Figure 7.10b). The system is grouped around a centre of symmetry and is planar. Furthermore, hydrogen bonds between the carboxyl groups also bind D–L pairs together in columns which are united to sheets by the OH hydrogen bond squares. In contrast,

---

[15] P. Luger, C. André, unpublished
[16] C.W. Lehmann, J. Buschmann, P. Luger, C. Demoulin, J.-H. Fuhrhop, K. Eichborn, *Acta Cryst.*, **1990**, *B46*, 646
[17] G.S. Parry, *Acta Cryst.*, **1951**, *4*, 131

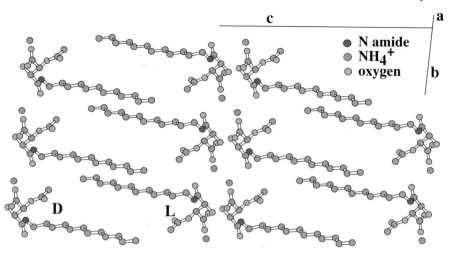

Figure 7.9 *Crystal structure of the monoammonium salt N-dodecyl-D,L-tartaric amide-2-acetate. The structure consists of interdigitated, racemic bilayers. The methyl group of an acetyl side-chain is bent in and connects the head group of one molecule with the alkyl chain of another.*

crystals of pure enantiomers of tartaric acids have no such hydrogen bond cycles[18]. As a result, the racemate is much less soluble than the enantiomers! 100 mL of water dissolve 133 g of L-tartaric acid but only 10.6 g of D,L-tartaric acid (= racemic acid). Only in very rare, exceptional cases can pure enantiomers be crystallized from racemic mixtures (spontaneous racemate resolution). Again, tartaric acid provides the most prominent example: below 30° the racemate of its ammonium sodium double salt separates into two different crystals with mirror image shapes (L. Pasteur, 1848).

Replacement of the carboxyl group of fatty acids by phosphate allows the formation of mono- and diesters without eliminating the negative charge. The chemistry of natural membranes is dominated by phosphodiesters containing a glyceride with two fatty acid esters and a polar ester with an amino alcohol, an amino acid or a carbohydrate. As an example, we chose the lecithin model compound dimyristyl-glycero-phosphate (DMP, Figure 7.11)[19,20]. DMP is electronegative and the anionic phosphate end bends towards the fatty ester groups. Both fatty acid chains are in the *all-trans* configuration. In this case, the chiral glyceryl carbon atom C2 is *S*-configured.

Lipids with two identical fatty acids such as DMP very rarely appear in natural membranes. Natural lecithins are mixtures of compounds which usually contain saturated fatty acids with chain lengths from C12 to C20 at carbon atom 1 and saturated *or* unsaturated, *cis*-configured fatty acids at the chiral carbon atom 2. Such mixtures of lecithins never form liquid crystals in mem-

[18] Y. Okaya, N.R. Stemple, M.I. Kay, *Acta Cryst.*, **1966**, *21*, 237
[19] K. Harlos, H. Eibl, I. Pascher, S. Sundell, *Chem. Phys. Lipids*, **1984**, *34*, 115
[20] I. Pascher, M. Lundmark, P.G. Nyholm, S. Sundell, *Biochim. Biophys. Acta*, **1984**, *1113*, 339

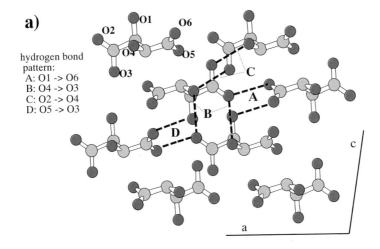

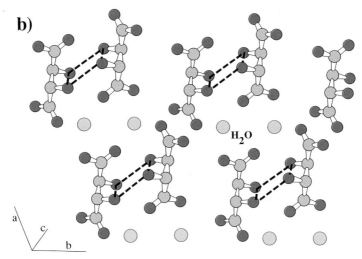

Figure 7.10 *Crystal structures of (a) D-tartaric acids, with no hydrogen bond between hydroxyl groups. Only hydroxyl–carboxyl hydrogen bonds are formed. (b) D,L-tartaric acid (= racemic acid) crystals contain strong hydrogen bond cycles between the hydroxyl groups of opposite enantiomers. The racemate is therefore much less soluble in water than the pure enantiomer[16].*

branes or 3D crystals. Both the multitude of chain lengths and flexibility of unsaturated fatty acids help to maintain natural membranes in a liquid state. Order in biological membranes only occurs in proteins and their environment. Crystalline membrane structures are the best defined and most interesting synkinetic molecular assemblies, but they play no role in nature.

Stereochemically, the most diversified synkinons are the eight diastereomeric

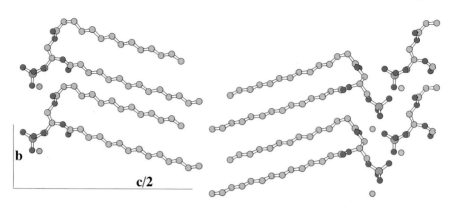

Figure 7.11 *Crystal structure of sodium 2,3-dimyristoyl-D-glycero-1-phosphate. An example of a chiral phospholipid with the typical bends of the ester groups at 0–1 and 0–3*[19,20].

N-alkylaldonamides and their enantiomers **1–10**[21]. Of these N-octyl-D-galactonamide is the least soluble ($\sim$ 16 mmol L^{-1} at 100 °C); the racemate D,L-galactonamide is even less soluble ($\sim$ 10mmol L^{-1}); and the D-mannonamide is ten times more soluble ($\sim$ 98 mmol L^{-1} at 100 °C), although the stereochemical interactions in both diastereomers are very similar. No 1,3-*synaxial* (= *ecliptic*) interaction of the hydroxyl groups occurs in the *all-trans* conformation. In glucon-, gulon- and talonamides, one such 1,3-synaxial interaction is found in the stretched conformation. The solubilities in boiling water increase by a factor of one hundred and more (glucon: 1.6 mol L^{-1}; talon 2.3 mol L^{-1}; gulon > 3 mol L^{-1}).

However, the critical micellar concentrations are barely influenced by solubilities[22]. They are close to 3 mmol L^{-1} in the cases of N-octyl-D-mannon-, D-glucon-, D,L-glucon- and D-gulonamides at 90 °C. Upon cooling to room temperature, all four diastereomeric aldonamides as well as the D-talonamide become insoluble (< 1 mmol L^{-1}). This is not the case for the N-octylamides of D-allonic, D-altronic and D-idonic acids. These diastereomers show two 1,3-*synaxial* interactions and are freely soluble at room temperature. One may therefore assume that the undisturbed linear conformation in galacton- and mannonamides favour the formation of intermolecular hydroxyl and amide hydrogen bonds, whereas unstable coiled conformations in allon-, altron- and idonamides prevent such interactions. Solubilities are thus inversely related to the rigidity of the open-chain carbohydrate head groups.

Stereochemistry does not only determine solubilities, but also steers the arrangement of amphiphilic molecules within a crystal as well as their relative

---

[21] J.-H. Fuhrhop, P. Schnieder, E. Boekema, W. Helfrich, *J. Am. Chem. Soc.*, **1988**, *110*, 2861
[22] J.-H. Fuhrhop, S. Svenson, C. Böttcher, E. Rössler, H.-M. Vieth, *J. Am. Chem. Soc.*, **1990**, *112*, 4307

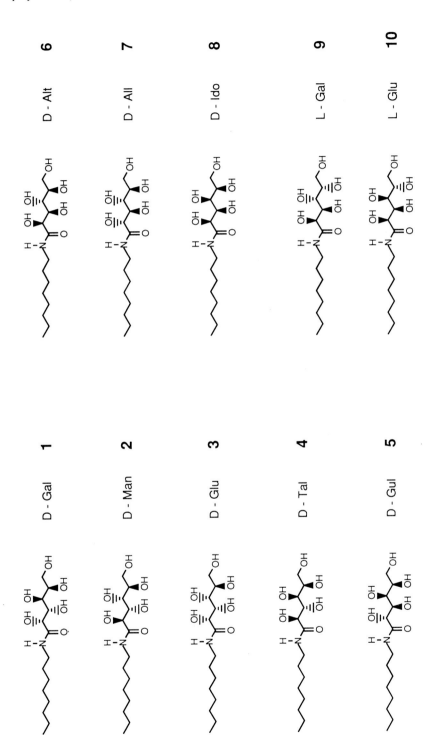

arrangement. The example of closely related *N*-octyl-D-glucon-[23] and D-gulonamides[24] is given here. *N*-octyl-D-gulonamide has the hydroxyl groups on C3 and C5 in *ecliptic* positions if one assumes the fully staggered *all-trans* conformation; gluconamide is in the same situation at C2 and C4. The crystal structure of both glyconamides shows a practically linear conformation of both diastereomers (Figure 7.12). Steric repulsion can thus be neglected in the water-free solid state, with intramolecular hydrogen bonding overcoming steric repulsion.

In the gluconamide, a *homodromic* hydrogen bond cycle was found in which four short hydrogen bonds run unidirectionally from one crystal plane to the other (Figure 7.12a). This kind of cycle is impossible with any hydroxyl group pair in the gulonamide crystal (Figure 7.12b). Inversion of the spatial orientations at C3 and C4 destroys the *homodromic* cycle. The strong hydrogen bonding cycle in gluconamide bends the terminal OH group towards the crystal sheet. Its surface is formed by a hydrophobic methylene group, whereas the gulonamide has a hydrophilic OH group. This difference has a dramatic effect on the crystal sheet orientation: gulonamide crystals remain in the expected (and most common) tail-to-tail orientation (Figure 7.12b), whereas the gluconamide sheets turn around to produce a head-to-tail orientation (Figure 7.12a). This is analogous to the observation that fatty acids crystallize in a tail-to-tail fashion, but their esters often head-to-tail (Figure 7.5). The dipolar interaction between alcohol groups is obviously much weaker than that between methoxy carbonyl groups.

The head-to-tail arrangement of the sheets in D-gluconamide crystals is again reverted to a tail-to-tail ordering in racemic D,L-gluconamide, as evidenced from TEM and X-ray data of microcrystals. The *homodromic* hydrogen bond cycle cannot be realized in the racemate pair so that the terminal CH_2OH group is presumably not bent in.

In the crystal structure of the bis-chiral bolaamphiphile octamethylene-1,8-bis-D-gluconamide **11**, the hydroxy groups of both end groups occur on the same side of the hydrocarbon chain[25]. This has been made possible by two *gauche* conformations of the connecting octamethylene chain. As a result, both end groups are connected in very similar homodromic hydrogen bond cycles within the crystal sheet on the same side of the chain's extension. This arrangement is as close as possible to the arrangement in the head-to-tail oriented sheets of the gluconamide amphiphile.

---

[23] a) V. Zabel, A. Müller-Fahrnow, R. Hilgenfeld, W. Saenger, B. Pfannemüller, V. Enkelmann, W. Welte, *Chem. Phys. Lipids*, **1986**, *39*, 313
b) C. André, P. Luger, S. Svenson, J.-H. Fuhrhop, *Carbohydr. Res.*, **1992**, *114*, 4159

[24] P.C. Moews, J.R. Knox, *J. Am. Chem. Soc.*, **1976**, *98*, 6628

[25] A. Müller-Fahrnow, W. Saenger, D. Fritsch, P. Schnieder, J.-H. Fuhrhop, *Carbohydr. Res.*, **1993**, *242*, 11

Amphiphilic Crystals and Hydrogen Bonded Co-Crystals

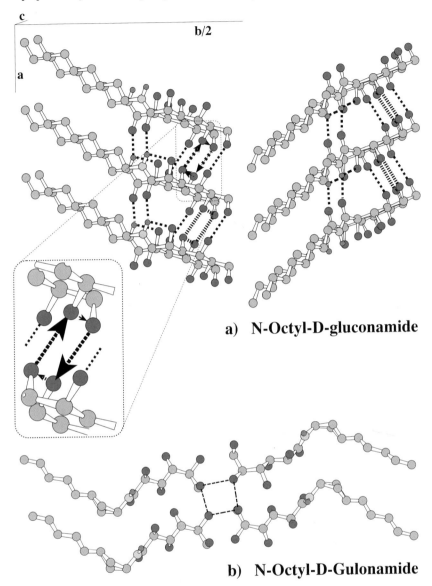

a) **N-Octyl-D-gluconamide**

b) **N-Octyl-D-Gulonamide**

Figure 7.12 *Crystal structures of (a) N-octyl-D-gluconamide (head-to-tail; homodromic intralayer hydrogen bond cycle) and (b) N-octyl-D-gulonamide (tail-to-tail; interlayer hydrogen bonds).*[23]

Crystal structures of glycolipids with a cyclic head group, e.g. 1-decyl-α-D-glucopyranoside, show an interdigitated bilayer. Two alkyl chains lie practically perpendicular to the carbohydrate plane. Interdigitated bilayers in crystals produce the hydrophilic interactions between the head groups typical for bilayer structures; the thickness of the interdigitated bilayer, however, is very similar to the thickness of a monolayer (Figure 7.13).

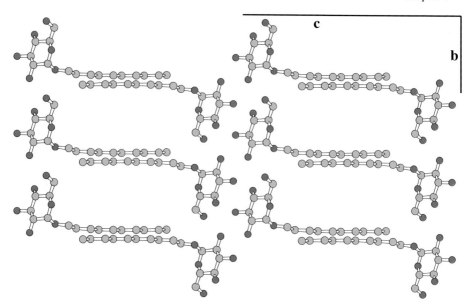

Figure 7.13 *Crystal structure of 1-decyl-α-D-glucopyranoside. The cyclic head groups trigger an interdigitation of the alkyl chains*[24].

7.5 Single Crystals of Carotenoids and Porphyrins

To a first approximation, biological cell membranes are either colourless or colorized by red porphyrins, green chlorophylls or yellow carotenes. These dyes render the membranes photo- and redox-active and are responsible for the primary reactions in photosynthetic charge separations and light-driven proton and metal ion transport in membranes as well as for the oxygenation of organic substrates. Integration of these molecules into synkinetic systems is of central significance. In the following, we introduce some molecular and crystal structures of linear carotenoids and macrocyclic porphyrins. Although many of these rigid, hydrophobic molecules occur as amphiphiles with carboxylate end groups (e.g. bixin, protoporphyrin IX) the crystallization of one of these amphiphiles has not yet been achieved. In the case of porphyrins with a hydrophilic edge, it was demonstrated that micellar fibres are very quickly formed but have a very low tendency to rearrange to planar crystal sheets (see section 5.8); so we therefore have to restrict ourselves to crystal structures of non-amphiphilic carotenes and porphyrins.

Carotenes are most *all-trans* configured polyenes with methyl groups in 1,5-positions. In crystals, the linear chains interdigitate with their non-methylated edges, whereas the methyl groups occupy the space above or below neighbouring molecules (Figure 7.14)[26]. No crystal structure of a *cis*-configured carotene (bixin, retinal) is known, only an *all-trans* retinal derivative was

[26] J. Hjortas, *Acta Cryst.*, **1972**, *B28*, 2252

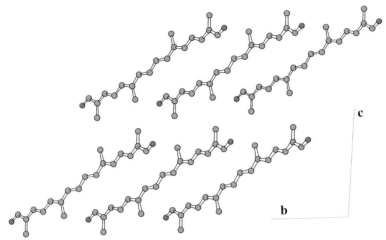

Figure 7.14 *Crystal structure of* all-trans *crocetine aldehyde with two terminal formyl groups. The methyl groups determine the packing arrangements*[26].

solved[27]. Carotene packing is, in general, dominated by sterical repulsion between the methyl groups (see section 5.7).

Comparing many porphyrin crystal structures[28], it turns out that the porphyrins are always parallel and no rotation of one porphyrin macrocycle relative to another occurs. Furthermore, electrostatic interactions steered the electron-rich pyrrole ring of one porphyrin over the electron-poor π-cavity at the centre of the neighbouring porphyrin (Figure 7.15). Such an arrangement minimizes π–π repulsion and at the same time, maximizes attraction of the s-framework around the inner edge of the π-cavity of one porphyrin with the π-electrons of the pyrrole ring immediately above[29]. The length and width of a porphyrin macrocycle is about 7 Å, its thickness 4 Å and the central hole about 2 Å wide (see Figure 5.20).

Meso-tetraphenylporphyrin (TPP) is an artificial porphyrin with four bulky benzene rings on the methine bridges. These rings' size and electron density are about the same as those of the pyrrole rings of the porphyrin and they lie perpendicular to the porphyrin plane (Figure 7.16a). This situation changes at acid pH values when the diprotonated dication is formed: the charge repulsion in the central cavity leads to a ruffled porphyrin macrocycle and the phenyl rings turn around (Figure 7.16b)[31]. The most important effect of the phenyl rings is the total hindrance of side-to-side interactions between porphyrin macrocycles and large stacking distances are enforced (see co-crystals in Figures 7.26 and 7.27). Electronic interaction between TPPs is minimal. These

[27] C.J. Simmons, A.E. Asato, R.S.H. Liu, *Acta Cryst.*, **1986**, *C42*, 711
[28] W.R. Scheidt, Y.J. Lee, *Struct. Bonding (Berlin)*, **1987**, *64*, 1
[29] W.S. Caughey, J.A. Albers, *J. Am. Chem. Soc.*, **1977**, *99*, 6639
[30] J.-H. Fuhrhop, *Angew. Chem.*, **1976**, *88*, 704; *Angew. Chem., Int. Ed. Engl.*, **1976**, *15*, 648
[31] E.B. Fleischer, *Acc. Chem. Res.*, **1970**, *3*, 165

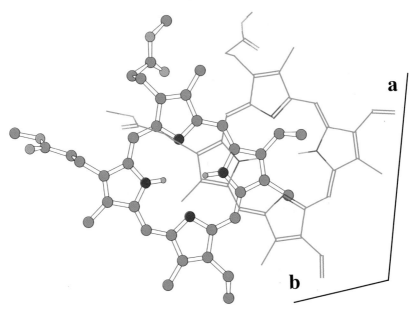

Figure 7.15 *The typical arrangement of a porphyrin dimer is exemplified here with the crystal structure of protoporphyrin IX dimethyl ester[29]. One pyrrole ring lies above the central cavity of the partner molecule.*

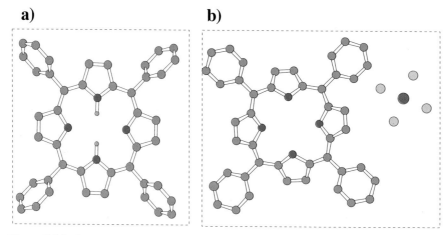

Figure 7.16 *In contrast to β-alkyl substituted porphyrins found in nature, the artificial meso-tetraphenylporphyrins do not form stable dimers. In crystal structures the molecules are well separated by the phenyl rings (see Figures 7.26 and 7.27) which are oriented perpendicular to the porphyrin macrocycle (a). Upon protonation with HCl and $FeCl_3$, the diprotonated porphyrin dication is formed. In its crystal structure the porphyrin ring is puckered and the phenyl rings are approximately in the porphyrin plane (b)[31]. Aggregation can now occur.*

porphyrins therefore produce the narrowest and most intense Soret band ($\Delta\lambda_{1/2} = 15$ nm, $\epsilon = 4 \times 10^5$).

7.6 Single Crystals and Fibres of Amphiphilic Steroids

Steroidal amphiphiles crystallize in normal tail-to-tail bilayers as long as linear, *all-trans* configured steroids are involved. Cholesteryl-3-sulfate dihydrate, for example, forms a bilayer with exactly aligned molecules (not shown)[32].

The same situation can also be found in arch-shaped or concave deoxycholic acid (= DC; Figure 7.17) derivatives, in which rings A and B are *cis*-fused. As an example, a bilayer was observed in monoclinic crystals of the rubidium salt of DC (= RbDC). **Hexagonal crystals of NaDC and RbDC hydrates**, on the other hand, produced interesting helical structures (Figure 7.17) with hydrophilic centres[33]. They **constitute the prototype of a well defined micellar structure,**

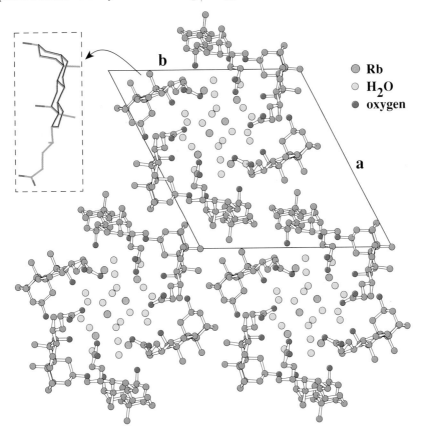

Figure 7.17 *Crystal structure of rubidium-deoxycholate, containing 10 water molecules per three deoxycholate molecules. The steroid molecules form water filled helices*[33] *(compare with Figure 3.11).*

[32] B.M. Craven, in ref. 4, p. 149ff
[33] G. Conte, R. Di Blasi, E. Giglio, A. Parretta, M.V. Pavel, *J. Phys. Chem.*, **1984**, *88*, 5720

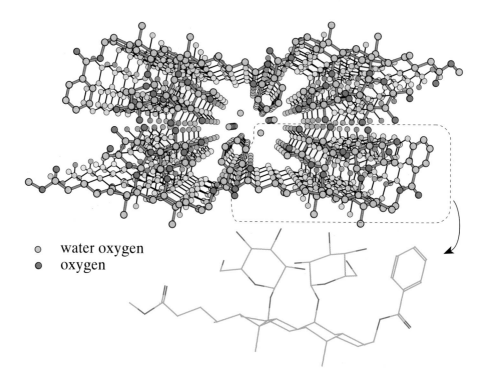

Figure 7.18 *Crystal structure of cholic acid diglucoside[34].*

in which curvature is determined not by the hydrophobic effect, but by the molecular shape of the monomers and strong intermolecular binding. These actions are due to the fact that DC is an edge amphiphile possessing two OH groups in α-positions and a carboxylic acid side-chain which also turns inwards.

The edge amphiphile character is even more pronounced in the glycosylated derivative of cholic acid[34] which crystallizes in the form of planar bilayers with the axial carbohydrate moieties in the centre. Large channels containing organized water molecules and potassium counter-ions are thus formed (Figure 7.18).

7.7 Molecular Recognition at Crystal Interfaces

Non-chiral molecules can possibly form chiral crystals. If the monomers of these chiral crystals are photoreactive, the supramolecular chirality may be utilized in order to induce molecular chirality in the photoproducts. One well established system is that of the aromatic diene given in Figure 7.19 which

[34] Y. Cheng, D.M. Ho, R. Gottlieb, D. Kahne, *J. Am. Chem. Soc.*, **1992**, *114*, 7319

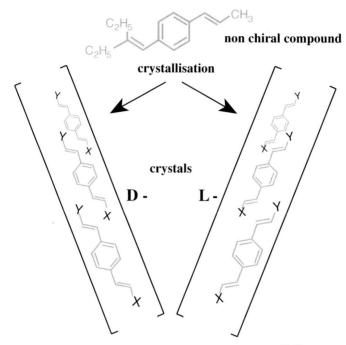

Figure 7.19 *Chiral crystals may form from achiral molecules*[35–37].

reacts to give pure enantiomers of substituted cyclobutadienes after the UV irradiation of one isolated crystal[35–37].

Similarly, crystals of achiral glycine differentiate between *R*- and *S*-amino derivatives. In the stable α-form of the glycine crystal the molecules adopt a conformation in which the C—H bond, which forms the CH···O contact, is parallel to the b-axis (Figure 7.20). One pair of these CH groups forms the (010) and (0$\bar{1}$0) crystal faces. The (*pro-R*) C—H bond is directed towards + b and turns away from the crystal bulk. The (*pro-S*) C—H bond is directed towards − b and thus into these crystals. Such an arrangement is named "enantiopolar". An α-amino acid additive of configuration *R* is capable of replacing a glycine molecule thus blocking crystal growth only at the (*pro-R*) sites away from the crystal. The growth along the + b direction is then hindered, leading to an increase in the area of the (010) face. Via symmetry an *S* amino acid practises the same effect on the (0$\bar{1}$0) face. The symmetric, bipyramidal crystals of α-glycine therefore changed to two asymmetric, mirror-image pyramids. Addition of racemic *R,S*-Ala suppresses growth in both − b and + b directions and a platelet with large (010) and (0$\bar{1}$0) faces forms. Then "engineered" crystal[38] not

---

35 L. Addadi, J. von Mil, M. Lahav, *J. Am. Chem. Soc.*, **1982**, *104*, 3422
36 J. Van Mil, L. Addadi, E. Gati, M. Lahav, *J. Am. Chem. Soc.*, **1982**, *104*, 3429
37 L. Addadi, Z. Berkovitch-Yellin, I. Weissbuch, J. Van Mil, L.J.W. Shimon, M. Lahav, L. Leiserowitz, *Angew. Chem.*, **1985**, *97*, 476; *Angew. Chem., Int. Ed. Engl.*, **1985**, *24*, 466
38 I. Weissbuch, L. Addadi, Z. Berkovitch-Yellin, E. Gati, S. Weinstein, M. Lahav, L. Leiserowitz, *J. Am. Chem. Soc.*, **1983**, *105*, 6615

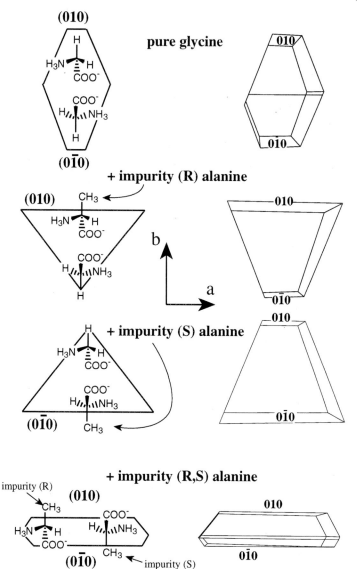

Figure 7.20 R-Ala adds to the (010) face and prevents the growth of the glycine crystal along the +b-axis. The (010) face becomes dominant because growth along the other faces is faster now. S-Ala has the same effect in the −b direction; R,S-Ala in ±b directions[35–37]. Enantiopolar crystals of achiral glycine add S-alanine to pro-S crystal faces and R-alanine to pro-R faces[37,38].

only changes its habit, but a more or less isolated layer rich in the Ala-impurity is also included.

The glycine platelets (Figure 7.20) have very large (010) and (0$\bar{1}$0) hydrophobic planes which swim in water at the surface with the b-axis oriented perpendicularly to the air/water interface. Either the (010) or (0$\bar{1}$0) face points to the bulk water phase allowing growth. If an (R,S)-amino acid mixture is dissolved in water, the crystals with the (010) face in water gather together the R-amino acid molecules and the crystals with the $(0\bar{1}0)$ face in water select the S-amino acid[39]. When the same glycine platelets swim on the solution of a pure enantiomer of a hydrophobic amino acid, e.g. S-leucine or S-phenylalanine, all pyramidal platelets turn their (010) faces to the air. These directed platelets can then be applied towards the resolution of concentrated solutions of further hydrophilic amino acids, for example R,S-glutamic acid[39].

Platelike crystals of chiral amphiphiles sometimes show opposite hydrophilic and hydrophobic faces if the amphiphiles are oriented head-to-tail. This can be detected by wettability measurements. In the case of N-octyl-D-gluconamide crystals (see section 5.6), the (010) face was hydrophobic and the $(0\bar{1}0)$ face hydrophilic. In combination with the crystal structure, the absolute configuration of the chiral centres can be determined and relative growth rates of hydrophobic and hydrophilic faces evaluated[40,41].

7.8 Crystal and Co-Crystal Engineering by Hydrogen Bonding and by the Filling of Voids

The major problem in the synkinesis of functional molecular assemblies lies in the co-crystallization of well-organized lipid mono- and bilayers with rigid, planar dyes. Biologically[42] and in model studies[43], this problem is solved by wrapping the dyes in several equally rigid protein helices which themselves can be integrated into membranes as edge amphiphiles. In synthetic lipid membranes without proteins, one must rely on co-crystallization processes, which ask for a perfect fitting of head groups and some compatibility of the hydrophobic parts. Synkinetic planning or "retro-synkinesis" can therefore be based on experience with hydrogen bonded 3D co-crystals. In the following section, a few recent examples are discussed.

The simplest networks are **one-dimensional α-networks** which may be composed of secondary amides, primary amide dimers or nucleophospholipids. In chapter 5, such structures were discussed as micellar rods and tubules in bulk aqueous solutions. Two-dimensional materials such as copper oxide superconductors, molybdenum sulfide lubricants and intercalated graphites are mostly inorganic. The anisotropic properties are a result of covalent bonds in two dimensions and weak interactions in the third dimension. One may, however, also envision strong hydrogen-bond interactions within an organic layer, whereas adjacent layers are held together only by van de Waals interactions. The two-dimensional, or β-network may form spontaneously from an

---

[39] I. Weissbuch, L. Addadi, L. Leiserowitz, M. Lahav, *J. Am. Chem. Soc.*, **1988**, *110*, 561
[40] J.-L. Wang, M. Lahav, L. Leiserowitz, *Angew. Chem.*, **1991**, *103*, 698; *Angew. Chem., Int. Ed. Engl.*, **1991**, *30*, 696
[41] J.-L. Wang, L. Leiserowitz, M. Lahav, *J. Phys. Chem.*, **1992**, *96*, 15
[42] R.E. Dickerson, I. Geis, Hemoglobin, Benjamin, Menlo Park, **1983**, p. 29ff
[43] C.T. Choma, J.D. Lear, M.J. Nelson, P.L. Dutton, D.E. Robertson, W.F. de Grado, *J. Am. Chem. Soc.*, **1994**, *116*, 856

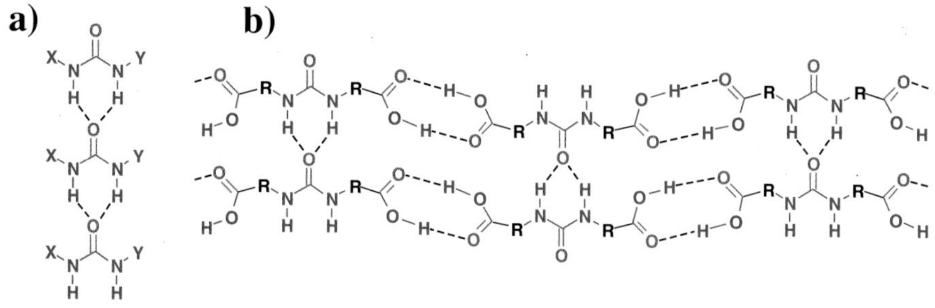

Figure 7.21 *a)* N,N'-*disubstituted ureas form α-networks by forming pairs of parallel hydrogen bonds. b) To produce a β-network a second hydrogen bond functionality must be introduced*[44].

α-network if the latter has P2 symmetry, meaning a two-fold axis parallel to the main axis[44]. *N,N'*-Disubstituted ureas, for example, form α-networks by forming parts of complementary hydrogen bonds only along one axis (Figure 7.21a). **In order to produce a β-network, a second hydrogen bond functionality at the edges (e.g. acids or amides) must be introduced** (Figure 7.21b). If one considers asymmetric molecules, an abundance of molecular packings become available. For example, an L-phenylalanine derivative of urea forms a β-network in which α-networks are linked by *like*-to-*unlike* amide to acid hydrogen bonds[44] (not shown). The racemate does not form β-networks[44].

The first attempt towards **three-dimensional diamondoid, tetrahedral networks** was made with adamantane-1,3,5,7-tetracarboxylic acid. Interpenetration of five different networks, however, precluded the formation of internal chambers. A more open structure with large chambers was obtained from the tetra-2-pyridone methane called "tecton". Its diamondoid lattice entrapped butyric acid (Figure 7.22)[45].

The classical case of co-crystals is that of weakly bonded electron donor–acceptor pairs. In crystalline 1:1 charge transfer complexes, donor and acceptor molecules alternate and lie flat, one on top of each other. The influence of charge transfer on crystal structures is at its greatest when the donor and acceptor are shifted against each other by half a molecular diameter[46]. The strongest connecting and ordering force is the formation of hydrogen donor–acceptor pairs. One such motif for solid-state molecular assemblies is the complementarity of dicarboxylic acids with bis(amidopyridines) containing aromatic spacers. Co-crystallization of a biphenyl derivative with 1,12-dodecanedicarboxylic acid, for example, forms structures with alternating arrangements of diacid and diamide (Figure 7.23)[47]. Here and in similar

---

[44] Y-L. Chang, M.-A. West, F.W. Fowler, J.W. Lauber, *J. Am. Chem. Soc.*, **1993**, *115*, 5991
[45] M. Simard, D. Su, J.D. Wuest, *J. Am. Chem. Soc.*, **1991**, *113*, 4696
[46] C.K. Prout, J.D. Wright, *Angew. Chem.*, **1968**, *80*, 688; *Angew. Chem., Int. Ed. Engl.*, **1968**, *7*, 659
[47] F. Garcia-Tellado, S.J. Geib, S. Goswami, A.Z.D. Hamilton, *J. Am. Chem. Soc.*, **1991**, *113*, 9265

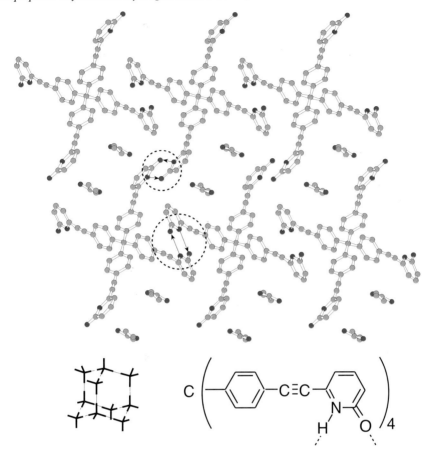

Figure 7.22 *The rigid tetrapyridonyl methane called "tecton" forms infinite diamondoid networks, which entrap butyric acid in crystals[45].*

co-crystals, a **perfect matching of long oligomethylene chains and planar dye molecules** dominates the crystal sheet arrangements.

Another co-crystallization involved the combination of a small isophthaloyl-aminopyridine receptor with a biphenyl diacid which, due to its size and conformational restrictions, could not hydrogen-bond to both aminopyridine sites. The crystal took up a cyclic arrangement of alternating diacid–diamide units linked by eight hydrogen bonds[48]. The overall shape became a figure-of-eight. No hydrogen bonds are formed between adjacent aggregates (Figure 7.24); curvature is thus introduced into the crystalline arrangement of planar hydrogen donors and acceptors.

Reaction of cyanuric acid (CA) with melamine (M) forms an insoluble 1:1 complex. This indicates directed hydrogen bonds between CA and M[49].

[48] J. Yang, E. Fan, J. Geib, A.D. Hamilton, *J. Am. Chem. Soc.*, **1993**, *115*, 5314
[49] C.T. Seto, G.M. Whitesides, *J. Am. Chem. Soc.*, **1990**, *112*, 6409

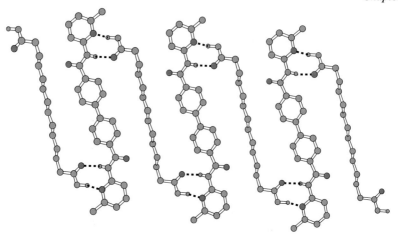

Figure 7.23 *Co-crystal structure of 4,4'-bis{[(6-methylpyrid-2-yl)amino]carbonyl}biphenyl and 1,12-dodecanedicarboxylic acid, a rare combination of extended sp^2- and sp^3-hybridized molecules[46].*

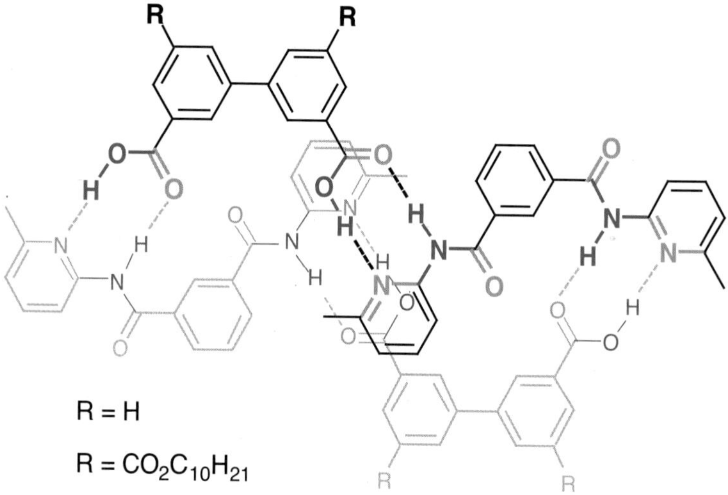

Figure 7.24 *Spontaneous assembly of a figure-of-eight by two planar molecules[48].*

Exchange of CA for a barbituric acid with alkyl substituents, e.g. diethyl in Veronal®, and substitution of two amino groups of melamine block the formation of β-sheets. If one assumes that a co-crystal retains the triad patterns of hydrogen bonds, then two tape structures and one cyclic structure remain plausible for the co-crystal (Figure 7.25). The linear tape was present with small X-groups (F, Cl, Br, I, CH_3); larger X-groups (CH_3) induced formation of the crinkled tape, whereas extra large tert-butyl groups led to the rosette[50]. Competition between nonbonded steric interactions in the R-groups ($= C_6H_4—X$)

[50] J.A. Zetkonski, C.T. Seto, G.M. Whitesides, *J. Am. Chem. Soc.*, **1992**, *114*, 5473

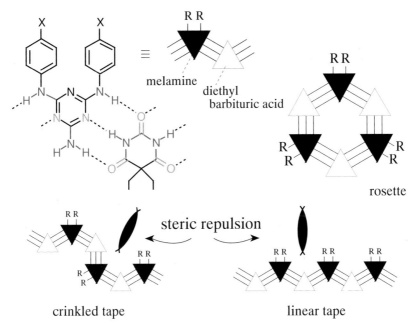

Figure 7.25 *Steric repulsion between a large substituent on melamine leads to curved assemblies between melamines and diethylbarbituric acid (= Veronal®10)*[49].

and a tendency for a high packing coefficient in the crystal are thought to be responsible for these structural motifs.

The design of lattice clathrates involves the use of blocks which are not entirely self-complementary; self-assembling blocks leave voids. **Tetraphenylporphyrin and its metal complexes are remarkably effective in producing microporous solids allowing the entrapment** of a series **of different molecules in co-crystals**[51–53]. In all these materials, sheets of porphyrins stacked to form parallel channels in which various guests were accommodated, e.g. naphthacenequinone (Figure 7.26). The minimum height of the channels corresponds to the thickness of the porphyrin core and is thus ideal for the incorporation of aromatic guests.

A mixed organic–inorganic 3D porphyrin network was obtained by co-crystallization of palladium *meso*-tetrapyridylporphyrinate and cadmium nitrate. All four pyridyl units are bound to octahedral cadmium centres which are each coordinated by two nitrate, water and pyridyl ligands (Figure 7.27). One half of the py–Cd–py connections is linear, the other half bent (103°), having two *cis* pyridine ligands. As is common in *meso*-tetraarylporphyrins,

[51] M.P. Byrn, C.J. Curtis, S.I. Khan, P.A. Sawin, R. Tsurumi, C.E. Strouse, *J. Am. Chem. Soc.*, **1990**, *112*, 1865
[52] M.P. Byrn, C.J. Curtis, I. Goldberg, Y. Hsion, S.I. Khan, P.A. Sawin, S.K. Tendick, C.E. Strouse, *J. Am. Chem. Soc.*, **1991**, *113*, 6549
[53] M.P. Byrn, C.J. Curtis, Y. Hsion, S.I. Khan, P.A. Sawin, S.K. Tendick, A. Terzis, C.E. Strouse, *J. Am. Chem. Soc.*, **1993**, *115*, 9480

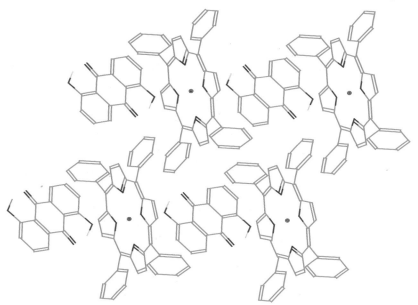

Figure 7.26 *Co-crystals of zinc* meso-*tetraphenylporphyrinate and 1,8-dihydroxyanthraquinone*[51–53]. *Such co-crystals cannot be obtained from natural* β-*octaalkylporphyrins (see Figure 7.15 and 7.16).*

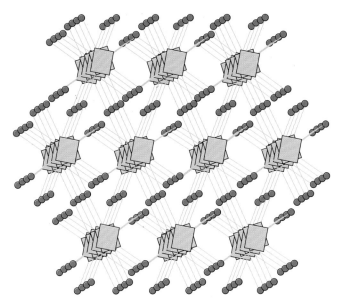

Figure 7.27 *Even inorganic salts, such as cadmium nitrate, can be entrapped in crystals of* meso-*tetrasubstituted porphyrins. Here one cadmium ion binds two pyridine rings from* meso-*tetrapyridinium porphyrin, two nitrate counterions and two water molecules (all symbolized by a blue spot)* [54].

[54] B.F. Abrahams, B.F. Hoskins, R. Robson, *J. Am. Chem. Soc.*, **1991**, *113*, 3606

steric clashes between the pyridines of one porphyrin and those of its face-to-face neighbours are avoided by staggering[54].

7.9 The Future of Synkinetic Membrane and Molecular Assembly Chemistry

The great diversity of concepts and synkinetic structures which have been realized within the last decade and which is partly represented in this volume, suggests that all kinds of membranes are accessible: asymmetric, as thin as 2.0 nm, helical, porous, fluid or solid, chiral on the surface or in the centre, photoreactive etc. etc. This diversity will inevitably grow. A few obvious unsolved problems which need immediate attention can also be detailed: e.g. synkinesis of solid micelles and vesicles from concave molecules with at least four hydrogen bonding sites, co-crystallization of porphyrins with solid membrane structures, and evaluation of nanopores as catalytic sites. Many more such target assemblies will undoubtedly be envisioned and successfully synkinetized.

What remains to be done is to integrate the ultrathin assemblies into society. This is being tried in many application labs throughout the world which work on LB surface monolayers. One may hope that they succeed in developing useful and stable devices. If this expectation is fulfilled, the whole field will flourish since (cast) monolayers, vesicles, micellar fibres and porous microcrystals can easily be combined and integrated. Much more is to be expected from building workable reaction systems or new materials than from mimicry of biological systems which has already been abandoned by most supramolecular chemists.

Subject Index

α-network, 207–208
Acetate ion, 96
Acetoacetate, 11
Acetylcholinesterase, 79
N6-Acetyl-9-propyl adenine, 30
Actin, 139
Adamantane-1,3,5,7-tetracarboxylic acid, 208
Adenine, 155
ADP = adenosine diphosphate, 155
Aerosol-OT (AOT), 43–44
AFM = atomic force microscopy, 30, 160
Agglutination, 63
Aggregation number, 24–28, 54
Alanine, 109, 205
Alignment, 104
Alkanenitrile, 161
Alkanethiol, 11–12, 19, 161, 164
Alkylation of amine, 8
All-trans conformation, 24, 121, 186, 187, 194, 198, 200
Allosteric transition, 137
Aluminium, 160
Amide, 104–106, 114, 147, 184, 208, 209
   aligned, 104
   cationic, 48
   D-glucon, 114
   dipole, 104
   entropy, 104
   free energy, 104
   glutamic acid, 106
   hydrolysis, 48
   hydrogen bond, 121
   hydrogen bonds, 104
   lack of hydrogen bonding, 193
   linearity, 104
   NMR, 105
   peptides, 104
Amidopyridine, 208

Amines, 11, 121
Amino acid, 22, 111
Ammonium ion, 186
Amorphous state, 160
AMP = adenosine monophosphate, 155
Amphiphile, 11, 17, 53, 131
Anthraquinone, 34, 179
Antitumor activity, 139
Antibody, 178
Antigene, 178
Arborol, 114
Artificial blood, 72
ATP = adenosine triphosphate, 155, 179
Azobenzene (and derivatives), 61, 79, 92, 108, 155, 190
Azonium group, 59

β-network, 207–208
Bacteriochlorophyll, 130
Barium sulfate, 155
Benzene dimer, 127
Benzenesulfinate, 161
Benzidine, 56
Benzoic acid, 27, 59
Benzyl radical, 93
Benzyltriphenylborate, 95
Bilirubin, 35
Biological membranes, 1, 7
Biopolymer fibres, 139–144
Biotin, 156
Biotinylated FAB, 157
Biphenyl, 209–210
Bipyramidal crystals, 205
Bipyridinium, 5, 9, 11, 15, 34, 36, 69, 72, 164, 178–179
Bixin, 10, 16, 200
BLM = bilayer lipid membrane, 1, 5, 50
   asymmetric, 59
   definition, 5

Subject Index

Bolaamphiphile, 8–19, 56–59, 76, 82–83, 86–88, 114
   asymmetric, 5–8, 14–15, 29, 37, 55–58
   asymmetric precipitation, 29, 37
   carotenoid, 15
   definition, 50
   ether linkage, 13, 82–83
   ferrocene, 17
   fluorinated, 15, 56–58
   half reduction, 37
   macrolides, 13–15, 51–57, 81–83
   quinone, 10, 16, 56, 70
   solubility, 50
   synthesis, 14–19
   vesicle membrane, 50–58
   zipper reaction, 12
Bond length, 187
Boranyl radical, 93
Borohydride, 75
Bouquet-shaped crown ether, 79
Bromide, 96
Bromo-undecanoyl chloride, 11, 17
Budding, 96
Butadiyne unit, 88
Butanol, 47

Calcium ion, 22
CAD (= DAC) = dodecylammonium chloride, 24
Cadmium, 176, 211–212
   porphyrin, 211–212
Calcein, 82, 85
Calcite, 152
Calixarene, 165, 168, 169
Carbohydrate recognition, 111–113, 118, 123–126, 147–148, 164–165
Carbonate, 151
Carbon tetrachloride, 165
Carbonyldiimidazole, 18
Carboxyfluorescein, 76
Carboxylate, 186
Caroviologen, 75
Carvone, 150
Cast films, 170, 172
Cation–proton transport, 80–81
CD = circular dichroism, 108, 126, 130, 133–137
Cerebroside sulfate, 56
Chain mobility, 51
Chain volume, 28

Channels 6, 74–75, 80–84, 168–169, 211
Chaotropic effect, 190
Charge transfer complex, 56, 208
Chelidamic acid, 61
Chiral amphiphiles, single crystals, 193
Chiral bilayer effect, 99, 120, 143
Chiral centre, 150
Chiral discrimination, 150
Chiral head group(s), 7, 13, 193
Chiral pool, 7, 1
Chirality, 118, 155
Chloride, 22, 63
Chloroform, 67, 165
Chlorophyll, 70, 135
Choleic acid, 34
Cholesteric phase, 147
Cholesterol, 54, 67, 76, 96, 147
Cholesterol oligoethylene glycol, 94
Cholic acid diglucoside, 204
Choline acrylate, 90
Chromatography, 7, 165
Chymotrypsin, 43
trans-Cinnamic acid, 32
Cis–trans isomerization, 92
Cloth-like structure, 121
Cluster, 22
Cmc = critical micellar concentration, 21, 52
Co-crystal, *see* crystal
Coated vesicle, 91
Collagen, 139–140
Computer modelling, 116–119, 133
ConA = concanavalin A, 63, 91
Conformational energy, 24
Conical shape, 28
Contact angle, 175, 177
Contour line diagram, 117
Copper, 33, 79, 160, 163, 177
   porphyrin, 30, 128, 133, 137
Covesicle, 59
Critical length, 28
Crocetin, 200
Crown-ether head groups, 61
Cryomicroscopy, 30, 103
Crystal, 3, 121–123, 150–155, 180–212
   2D crystal, 182–185
   3D crystal, 185–212
   achiral → chiral, 205
   alanine, 205–207
   azobenzene amphiphiles, 190

Crystal (*cont.*)
  bipyramidal, 205–207
  carotenoids, 200–201
  channel, 203–204
  chiral amphiphiles, 193–200
  chiral discrimination, 150, 184–185, 195–200
  cholesteryl-3-sulfate, 203
  cholic acid diglucoside, 204
  co-crystal, 143–144, 182–185, 195, 207–212
  crocetin, 201
  1-decyl-α-D-glucopyranoside, 199
  deoxycholic acid, 35, 203
  deoxycholic acid micelle, 35
  DMP, 194
  DODAB, 190
  enantiopolar, 205
  ethyl stearate, 188–189
  glycine, 205–207
  head-to-head arrangement, 193
  head-to-tail arrangement, 118, 187, 198
  helical arrangements, 35
  hydrophilic faces, 207
  hydrophobic faces, 207
  interlayer interaction, 186
  intralayer interaction, 186
  lecithin, 99
  linolenic acid, 190
  methyl stearate, 188–189
  molecular recognition, 204
  octamethylene-1,8-bis-D-gluconamide, 198
  N-octyl gluconamide, 187
  oleic acid, 190
  piperazinium bis-dodecanoate, 186–187
  porphyrin co-crystal, 211–212
  porphyrin amphiphiles, 131
  porphyrins, 200
  potassium palmitate, 187
  pro-R faces, 205
  protonated TPP, 201
  protoporphyrin IX dimethylester, 202
  racemic acid resolution, 194
  seeds, 123
  solid state NMR, 121
  stearic acid, 186
  steroid, 203
  tail-to-tail arrangement, 118, 187–190
  TPP, 201
  UV irradiation, 205
  zinc 5-pyridyl-10,15,20-triphenylporphyrin, 131
  zinc octaethyl porphyrinate radical dimer, 127
  zinc octaethylformylbiliverdinate, 145
CTAB = cetyltrimethylammonium bromide, 22, 24, 29, 30, 102
Curvature, 29, 99, 110, 122, 204
Cyanuric acid, 209
Cyclam, 61
Cyclic peptide, 141
Cyclic voltammetry, 179
Cyclodextrin, 140
Cyclohexane, 47
N-Cyclohexyl-*N'*-[β-(*N*-methylmorpholino) ethyl] carbodiimide-*p*-toluenesulfonate, 87
Cysteamine, 181
Cysteine, 151
Cystine, 151
Cytoskeleton, 89

Decyl-α-D-glucopyranoside, 199
Defect density, 167
Dendrimer, 39–43, 152, 183–184
  integrated porphyrin, 43
DC = deoxycholic acid, 34, 36, 203–204
Deoxynucleotide, 163
Deprotonation with acid, 68
Deuteration, 8, 27, 109
Dextrane, 101
Dextrin, 140
DHP = dihexadecyl phosphate, 70, 89
Diacetylene, 113
Dialysis, 75
Diamondoid network, 208
Diazirine, 94
Diazo dye (*see also* azobenzene), 155
Diazonium salt, 59
Diazotation, 11
Dicarbocyanine, 95
Dicyclohexylcarbodiimide (DCC), 18
Didodecyldimethylammonium bromide, 96
Dielectric constant, 22
Diffusion coefficient, 177
Dihexadecyldimethylammonium bromide, 61, 75, 79
Diketone complex, 163

Subject Index

Dimethylaminopyridine (DMAP), 18
Dimethyloctadecylsilyl silica, 165
Dimyristoyl-phosphatidyl-choline, 9
Dioxolane ring, 76
Dipalmitoyl-phosphatidyl-choline, 9
Dipalmitoyl-phosphatidyl-serine, 9
Dipalmitoyl-sn-glycerol-1-phospho-3-azidothymidine, 65
DIPEP = bis-phosphatidylethanolamine (trifluormethyl)phenyldiazirine, 93
Disks, 118
Distance effect, 164
Disulfide, 77, 87, 169
Dithionite, 75
Diyne, 18, 64–65, 88–89, 111, 113
DMMA = N-dimethylmyristamide, 135
DMP = dimyristoyl-phosphatidic-acid 9, 194
DMPC = dimyristoyl-phosphatidyl-choline, 9
DMSO = dimethyl sulfoxide, 133
DNA = deoxyribonucleic acid, (see also nucleic acid, polynucleotides), 85, 137, 139, 143–44
DODAB = dioctadecyldimethyl ammonium bromide, 52, 70, 190
DODAC = dioctadecyldimethyl ammonium chloride, 59, 77
Dodecanedicarboxylic acid, 208
Dodecylmalonic acid, 104
Dodecylimidazole, 72
Dodecyltrimethylammonium bromide, 190
Domain formation, 75–79
DPPC = dipalmitoyl-phosphatidyl-choline, 9, 56, 66, 75, 93, 150
DPPSer = dipalmitoyl-phosphatidyl-serine, 9
DSC = differential scanning calorimetry, 87
Dy = dysprosium, 85
Dynorphin A, 141

Edge amphiphile, 140, 204, 207
  definition, 4
Edge-to-edge porphyrin interactions, 133
EDTA = ethylenediaminetetraacetic acid, 45, 82–83
Efflux rate, 75

Eicosane sulfate, 29
Elastin, 139
Electrochemical detection, 149
Electrode surface, 38, 165–167
  micelle, 38
Electron acceptor, 134
Electron donor, 55, 134
Electron microscopy, ix, 53, 103, 106–108, 113–117, 130–132, 151–153, 160, 172
Electron transfer, 70, 178
Electrostatic interaction, 200
Ellipsometry, 175
Ellipticine, 139
Enantiomer, 99, 150
Enantiopolar arrangement, 205
Entrapped water volume, 85–86
Entropy, 21, 104
Epifluorescence microscopy, 151
Epitaxial matching, 154
EQCM = electrochemical quartz crystal microbalance, 163
ESR = electron spin resonance, 43, 72
Ester group, 151
Ether linkage, 12
Ethidium ion, 138
Ethyl group, 67
Ethyl stearate, 187
Ethyl xanthogenate, 11
Ethylenediamine, 12
Europium(III) ion, 146
Even–odd effect, 109, 161–162, 190
EXAF = extended X-ray absorption fine structure, 35
Excitons, 134
Extrusion, 99
Eye lens, 140

Facial amphiphile, 34
Fat droplet, 45
Fatty acid, 161, see crystal
FCCP = carbonyl cyanide 4-[(trifluoromethoxy)phenyl] hydrazone, 80
Ferricyanide, 33, 75, 167
Ferrocene, 11, 17, 30, 38–39
Fibrous assemblies, 3, 5, 98–143
  bacteriochlorophyll, 130
  bacteriopheophytin, 130
  bilayer rods, 114

Fibrous assemblies (*cont.*)
   bilayer rods of *N*-octyl-D-gluconamide, 114
   biopolymer, 139
   chiral bilayer effect, 120, 143
   cloths, 124
      N-dodecyl-D-tartaric amide, 124
   co-crystals, 123–126
   crystallization, 106
   cyclopeptide, 141
   D,L-polylysine, 143
   DNA, 137–139
   donor–acceptor interaction, 134
   exciton interaction, 134
   eye lens, 140
   fluid, 101–104
   gauche rotations, 109
   helical dimers, 144–145
   helical metal complexes, 144–147
   Jelley aggregates, 134
   linewidth narrowing, 134
   lyophilization, 99
   lysine, 110
   metalloporphyrin, 126–139
   micellar, 3, 98, 106–107, 114–126
   monolayer rods, 114
   N-octyl-D-gluconamide, ix, 117–126
   D,L-polylysine, 143
   polymerization, 111–113
   porphyrin, 126–139
   quadruple helix, ix, 117–119
   scrolls, 111
   solid, 106–126
   solid state NMR, 109
   synkinesis, 3
   vesicular, 98, 106–107, 110–113, 141–142
Finkelstein reaction, 11
Flash photolysis, 70
Flattening, 94
Flip-flop, 58–59, 62
Fluorescence, 6, 67, 70, 133–138, 152, 153, 177–178
Fluorinated amphiphiles, 13–15, 56, 58, 160
Formamide, 130
FPPR = photobleaching recovery, 177
Free energy, 104
Freeze fracture, 87
Fullerene, 75, 147, 173, 179–181

Future, 213

G^+ conformation, 121–123, 125
Gallium arsenide, 161
Gauche conformation, 24, 52, 121, 161, 187, 190, 198
Gel chromatography, 85, 131
Gel stability, 140
GID = grazing incidence diffraction, 182, 184–185
L-Glucose, 148
Glutamic acid, 13, 19, 143
N-Glutaryl-L-phenylalanine *p*-nitroanilide, 43
Glutathione, 70, 178
Glycerophospholipid, 8–9
Glycine, 205
Glycolipid, 65
Glyconamides, 13
Glycoprotein, 65, 139
Glycose tetraacetates, 63
Gold, 44, 160, 167, 175–181
Guanidinium amphiphile, 155

H-aggregate (hypsochromic shift), 61, 191
Haemagglutinin, 65
Haemoglobin, 45
Hairy rods, 173
Half-mustard, 47
Half-neutralization, 121
Harpoon, membrane-disrupting, 76
HD = hexadecane, 175
Head group area, 28, 53
Head-to-head arrangement, 96, 193
Head-to-tail arrangement, 118, 187, 198
Headgroup–headgroup repulsion, 96
Helical metal complexes, 144
Helix, ix, 114, 116–118, 141, 170
Heme fibre, 133
Hemepocket, 73
Heterodimer, 128–129
Heteroleptic compound, 129
Hexadecane, 175
Hexadecanethiol, 161, 177, 178
Hexadecylmethylamine, 11
Hexagonal pattern, 160
Hexameric helical bundle, 141
Homochiral arrangement of monolayers, 150
Homodimers, 127–128

Subject Index

Homodromic hydrogen bond cycle, 184, 198
Human immune deficiency virus (= HIV), 66
Hydration contour line, 22
Hydration force, 21
Hydrogen bond, 21, 208–212
　cluster, 22
　contour line, 22
Hydrophilic spacer, 11, 17, 87, 170
Hydrophobic edges, 118
Hydrophobic effect, 21
Hydrophobic face, 207
Hydrophobic gap, 26
Hydrosulfide, 11, 13, 19, 169
Hydroxy stearic acid, 4
Hydroxybenzoic acid, 102
Hydroxythiophenol, 167
Hyperextended amphiphile, 38
Hypochlorite, 47

Image analysis, 117
Immunoglobulin, 157
Indigo, 75
Influenza virus, 65
Infrared spectra, 121
Inhibitor, 65
Inner surfaces, crystals, 182
Integral porphyrin, 69
Intercalation, 137
Interdigitation, 187, 190, 193, 199
Interlayer interactions, 186
Intralayer interactions, 186
Inverted DNA, 146
Inverted micelle, 43
Iodide, 55, 190
o-Iodosobenzoate, 77
Iodosulfonic acid, 15
Iodophenol, 102
Ion binding, 163
Ion pair, 72
Ion transport, 75
Ionene polymer, 87
IR = infrared spectrum, 129, 161, 179
Iridium, 147
Iron(II) ions, 82
Iron porphyrin, 72, 136
Irreversible photochromism, 84
Isocyanuric acid, 173
Isophthaloyl-aminopyridine, 209

Isoprenoid side-chain, 167
Isotherm, 151

J-aggregate = Jelley, 134, 191

Kemps's acid, 155
Kinetic nucleation, 154
Krafft point, 21

L-glucose, 148
Lactones, 6, 7, 12, 13
Lanthanide anion, 85
Laser desorption mass spectrometry, 161
Lattice clathrate, 211
Lattice parameter, 155
Lauroyl compounds, 186
LB = Langmuir–Blodgett film, 149
Leaky membrane, 76
Lecithin, 99, 101
Lectin, 63
Lithium ion, 22
Light microscopy, 96, 99, 106, 151
Linewidth narrowing, 134
Linolenic acid, 190
Lipase, 66
Liquid crystal, 109, 147–148
Lysine, 110

Macrocyclic tetraether, 12
Macrolide, 50
Magic angle, 118
Magnesium porphyrin, 72, 130
Marcus theory, 178
Mass spectrometry 161
Matching, 209
Mathematica® computer program, 118
Mean surface, 190
Mean volume, 190
Melamine, 209
Melanin, 173
Membrane proteins, 85
Membrane rupture, 92
Mercapto-β-D-glucose, 80
Mercaptosuccinic, 8, 14
Mercaptoundecanoyl-sn-glycero-3-phosphocholine, 87
Mercury, 161
Metachromatic effect, 56–57

Metallo-tetrasulfonatophthalocyaninate, 129
Metallochlorin, 129
Metalloporphyrin, 126
Methacrylate, 12, 18, 91, 170
Methine cyanine, 134
Methyl group, 67
Methyl orange, 107
Methyl stearate, 187
Methylcyclohexane, 93
Methylene blue, 58
N-Methylmyristamide (MMA), 135
Magnesium ion, 22
Micelle, 2–5, 21–48, 99
   aggregation number 21–22, 26, 28
   anthraquinone, 34
   block model, 26
   centre, 28
   cmc = critical micellar concentration, 21–22, 34, 39
   computer simulation, 21, 27
   cone, 28
   conversion to vesicles, 36
   covalently dissolved molecule, 39
   cubes, 24
   disks, 118
   dissolution of molecules, 30
   dynamics, 29–30
   electrode surface, 38
   enthalpy, 21
   entropy, 21
   ferrocenyl, 38
   free energy, 21
   free enthalpy, 21
   hydration, 21
   hydration force, 21
   hydrogen bond, 21
   inverted, 43–44
   Krafft point, 21–22
   light induced electron transfer, 33–34
   lyophilization, 99
   neutron scattering, 27
   nucleic base pairs, 30–32
   phthalocyanine, 38
   pincushion, 26–27
   pK_a value, 27–30
   porphyrin, 30
   pyrene, 30
   quinone, 38
   reef model, 26–27
   SDS = sodium dodecyl sulfate, 22, 25, 27, 30, 32, 39, 93
   spheres, 24
   stability, 21
   steric force, 21
   steric repulsion, 21
   structural models, 24
   synkinesis, 2
   TEM = transmission electron microscopy, 29
   translucence, 36
   wetness, 27
   worm-like, 30, 102–104
   X-ray studies, 35
Michael addition, 8, 13
Microemulsion, 20, 44–48
Minimal surface, 118–119
Mixed anhydride, 10, 16
Mixed monolayer, 150–152
Mixed bilayer, 77–78
MLM = monolayer lipid membrane, 50–52, 54, 55–59
   definition, 5
Mobility, 51
Molecular harpoon, 76
Molecular recognition, 149–181, 204
Molecular shape, 28, 204
Molecular surface area, 149
Molecular wire, 73
Mono- and bilayer micellar rods, 114
Monolayers, 3–7, 149–181
   AB block polymer, 171
   amorphous state, 160
   antibody–antigene, 178
   avidin–biotin, 157
   bolaamphiphile, 173, 175–181
   Brewster angle light microscopy, 183
   calcite, 152
   calixarene, 165
   carbohydrate recognition, 165
   carboxylate ammonium bilayer, 178
   cast films, 172
   chiral discrimination, 150–153, 155, 165
   contact angle, 175, 177
   crystalline, 182
   crystalline state, 160
   defect, 167–169
   dendritic, 184
   dye, 175
   electrochemical detection, 149

Subject Index

Monolayers (*cont.*)
  electrostatic matching, 154
  ellipsometry, 175
  even–odd, 162
  fluorescence, 177
  fluorinated, 160, 184
  GID = grazing incidence diffraction, 182, 184–185
  glutathione, 178–179
  hairy rods, 173, 175
  helices, 170
  inorganic, 152
  ion binding, 163
  IR = infrared spectra, 175–176
  isoprenoid chain, 167
  melting, 160
  molecular recognition, 149–181
  molecular surface areas, 149
  multiple, 170–172
  nanopore, 149, 165
  photobleaching, 177
  photopolymerization, 171
  polymerization, 170–175
  pore, 168–169
  porphyrin, 179
  quartz crystal microbalance, 164
  quinone, 167
  racemic, 184
  receptor recognition, 155
  replacement, 169
  SAM = self assembling monolayer, 160, 164–181
  solid, 170
  stearate, 154
  stereochemical matching, 154
  steric repulsion, 164
  surface pressure isotherm, 150–161
  temperature dependance, 151–153
  triple layer, 157
  vaterite, 152
  viologen, 179
Monte Carlo simulation, 160, 168
Morphology, 154
Mosaic structures, 99
Multilayer, 170–172
Mustard, 47
Myelin figure, 99–100
Myristoyl compounds, 9, 135, 186

N6-acetyl-9-propyl adenine, 30

N-cyclohexyl-N'-[β-(N-methyl-morpholino)ethyl]carbodiimide-p-toluenesulfonate, 87
N-glutaryl-L-phenylalanine p-nitroanilide, 43
N-methylmyristamide (MMA), 135
N-octyl-D-allonamide, 196
N-octyl-D-altronamide, 196
N-octyl-D-galactonamide, 111–112, 196
N-octyl-D-gluconamide, 114–126, 183–184, 196–199
N-octyl-D-gulonamide, 196–197
N-octyl-D-idonamide, 196
N-octyl-D-mannonamide, 111–112, 123–128, 196
N-octyl-D-talonamide, 196
NaDC = sodium deoxycholate, 35–36
NaLS = sodium lauryl sulfate (*see also* SDS), 22, 24
Nanogram quantities, 164
Nanopore, 149, 165–169
Negative staining, 29, 119
Networks, α- and β-, 207–208
Neuroblastoma, 79
Neutron scattering, 27
Nickel porphyrin, 137
Nitrobenzoate, 59
Nitrogen ylide, 60
Nitrophenolate, 59
Nitrophenyl ester, 61
NMR = nuclear magnetic resonance, 27, 28, 32, 109, 141, 166, 186
  2D crystals, 186
  3D crystals, 186
  acetylenic protons, 27
  benzoic acid, 27
  ^{81}Br-NMR, 38
  ^{13}C T_1-values, 24
  cholestane nitroxide, 35
  conformation, 120
  CTAB fibres, 102
  deoxycholic acid, 35
  deuterium, 109, 166
  dynorphin A, 141
  fibres, 120–122, 186
  helix, 141
  in micelles, 32
  molecular shape, 24, 28
  nitrophenolate, 43
  ^{32}P-NMR, 51

NMR (*cont.*)
    shift reagent, 85
    silane monolayer
        high-loading, 166
        low-loading, 166
    solid state, 109, 120–122
        lyophilized fibres, 120–122
        signal assignment, 120–122
    solvent effects, 27
    tailored excitation pulse, 32
    vesicle, 51
Non-complementary polynucleotide, 156
Nonacosan-10-ol, 99–102
NTA = nitrilotriacetic acid, 85
Nucleic acid (*see also* DNA)
    bases, 32, 65, 111, 156
    helices, 143–144
Nucleotide head group, 65, 111

O-iodosobenzoate, 77
Octadecanethiolate, 160
Octanethiol, 13, 19
Octopus-porphyrin, 133
N-Octyl-D-allonamide, 196
N-Octyl-D-altronamide, 196
N-Octyl-D-galactonamide, 111–112, 196
N-Octyl-D-gluconamide, 114–126,
        183–184, 196–199
Octylglucoside, 86
N-Octyl-D-gulonamide, 196–197
N-Octyl-D-idonamide, 196
N-Octyl-D-mannonamide, 111–112,
        123–128, 196
N-Octyl-D-talonamide, 196
Odd-numbered hydrocarbons, 190
Oleic acid, 190
Oleoyl-lysophosphatidic acid, 9
Oligo-1,2-bipyridine, 144
Oligopeptide, 79
One-sided precipitation, 37, 55
Orientation effect, 22
Orthorhombic-to-hexagonal transition, 190
OT = dioctylsuccinate, 43–44
Oxidative bleaching, 45
Oxidative coupling, 11
Oxidative polymerization, 87
Oxygen, 72, 137
Oxygen adduct, 137

π-systems, 133

p-sulfonatophenyl, 139
Palmitate, 48, 186
    potassium, 187
Palmitic acid, 186
Palmitoyl lysine, 155
Palmitoyl-*R*-lysine, 184
Palmitoyl phospholipids, 9
Paramagnetic lanthanide anion, 85
Patch-clamp experiments, 142
PEG = polyethylene glycol, 82
Peptide(s), 43, 104, 140–143
    cyclic, 141–143
Perchlorate, 29, 55, 164, 190
Perfluorododecylaspartate, 184
Perfluorodecanol, 13
Permselectivity, 168
PET = photoinduced electron transfer, 93
pH gradient 77, 80
Phase transition, 66, 89
Phenothiazine, 33
Phenyl boronic acid, 147
Phenyl radical, 93
Phenylalanine, 207
Phenylenediamine, 9, 11, 15, 59, 73, 178
Pheophytin, 70
Phopholipid bilayer, 142
Phosphate, 121, 155
Phosphatidic acid, 56–58
Phosphatidylcholine, 56, 73
Phosphodiesters, 8
Phospholipase, 9, 66
Phospholipids, 8–9, 19, 77–79
Phosphonate, 155
Phosphotetrabromide anion, 173
Phosphotungstate, 110, 114
Photobleaching, 177
Photochromism, 84
Photodimer, 32
Photoionization, 33
Photolabelling, 94
Photopolymerization, 171
Photospiran, 84
Phthalocyanine, 128–129
Picket fence porphyrin, 72
Pincushion, 26–27
Pine needles, 99
Piperazinium bis-dodecanoate, 187
Place-holder, 169
Platinized cadmium sulfide, 44
Platinum, 43, 161

Subject Index

Polar activator, 134
Poly(1-vinyl-2-pyrrolidone), 137
Poly(A), 156
Poly(2-acrylamido-2-methyl-1-propane-sulfonic acid), 171
Polyacrylate, 45, 59
Poly-L-alanine, 170
Polyamine, 140
Polycation, 87–90
Poly[(dA-dT)$_2$], 137
Polyelectrolyte, 56
Polyene, 15–16, 74–75
Polyether, 140
Poly(2-ethylacrylic acid), 91
Poly-L-leucine, 170
Polymerization, 11, 17, 79–85, 108, 111, 170
Polymorphism, 187
Polynucleotides, 143–144, 156
Polyrotaxane, 140
Polysaccharide, 89, 120, 139–140
Polystyrene, 136
Polystyrene/poly-4-vinylpyridine AB block polymer, 171
Poly(U), 156
Polyurethane, 140
Poly(1-vinyl-2-pyrrolidone), 136
POP = 1-palmitoyl-ω-oleoyl-sn-glycero-3-phosphocholine, 77
Pores, 6, 74–75, 80–84, 168–169, 211
Porphyrin, 6, 15–16, 43, 66–74, 103, 126–139, 200–203, 211–213
   amphiphile, 15–16, 66–74, 126–139
   bacteriochlorophyll, 130
   bacteriopheophytin, 130
   cadmium, 211–212
   carbohydrate amides, 15–16, 131
   carbohydrate interaction, 136
   chlorophyll, 70, 129, 135
   co-crystal, 211–212
   copper, 30, 128, 133, 137
   copper fibre, 133
   crystal structure, 200–203, 211–212
   CTAB fibres, 103
   deprotonation, 68
   5,15-diphenyl-10,20-dipyridinium (= *trans*-DPyP), 138
   DNA adducts, 137
   DNA cleavage, 139
   edge-to-edge fibres (= I type), 133
   electron microscopy, 131
   ellipticine, 139
   ESR = electron spin resonance, 30
   face-to-face fibres (= H type), 133
   fibres, 126–133
   flash photolysis, 70
   haemoglobin model, 72, 136
   heterodimers, 128–129
   homodimers, 127–128
   integral, 69
   intercalation, 137
   ion pair, 72
   iron, 72, 136
   light-induced charge separation, 68
   magnesium, 72, 130
   metallochlorin, 129
   micelles, 30
   nickel, 137
   NMR, 129–130
   octaacetic acid, 68–69
   octaethyl (OEP), 127, 129
   octopus, 133
   oxygen, 72, 136
   packing, 127
   picket fence, 72
   pK_a value, 30
   polar activator, 134
   polymers, 136–139
   protoporphyrin, 131, 133, 134, 136
   quinone, 72
   radical dimers, 127–128
   stacking, 126–129, 132–134
   tetraphenyl (TPP), 72, 127, 128, 134, 200, 211
   tetrapyridinium, 128, 137
   tetrapyridyl 137, 211
   thorium(IV), 129
   tin(IV), 131–132
   triplet, 72
   uranium(IV), 129
   vanadium, 137
   vesicle, 72
   meso-viologen octaethyl, 69
   zinc, 72, 127, 128, 130, 133
Potassium ion, 81, 186
Potassium palmitate, 81, 187
Pyrene, 33
Pyridinium 190
Pyromellitic acid, 81

Quadruple helix, ix, 116–118, 141

Quartz crystal microbalance, 164
Quartz surfaces, 179–181
Quinone 6, 10–11, 16, 17, 56, 70, 72, 167

Racemate, 99, 123, 150, 184, 194, 208
Raman spectra, 24
Reductive cleavage, 8
Repulsion, 20, 26, 98, 118, 169, 198, 200
Retinal, 200
Reversible polymerization, 13, 19
Ripple, 66
Rosette, 210
Rotation of alkyl chain, 162, 190
Rubidium, 141, 203
Ruthenium(II), 34, 167

Salicylic acid, 102
SAM = self-assembled monolayer (*see also* monolayers), 160–164, 167, 177
SAXS = small-angle X-ray scattering, 35
Scanning tunnelling electron microscopy, 153
Scheibe polymers, 134
SDS = sodium dodecyl sulfate (*see also* NaLS), 22, 25, 27, 30, 32, 39, 47, 93, 122, 124
Second-harmonic generation, 179
Self-assembled monolayer, (*see also* monolayers), 160–164, 167, 177
Self-organization, 2
Semiconductor, 161
Serin, 151
Sexipyridine, 145
SHG = second-harmonic generation, 179
Sialic acid, 65
Sickle cell haemoglobin, 139
Signal molecule, 168
Silica 166
Silicon phthalocyanate, 173
Silicon surfaces, 179
Silver, 147, 175–177
Simulation, 160, 168
Sitosterol, 67
Skeletonized vesicle, 89
Sodium borohydride, 71
Sodium ion, 22, 45, 79, 104, 186
Solubility, 185, 194, 196
Solvophobic effect, 20
Soret band, 127, 131

Spacer arm, 172
Sphingophospholipid, 9
Spin label, 56
Spironolactone steroid, 167
SSSB = *N*-[(*m*-sulfobenzoyl)oxy] sulfosuccinimide, 93
Stacking, 126
Stationary phase, 165
Statistical lattice model, 24
Stearate anion, 154, 186
Stearic acid, 4, 186, 187
Stearoyl aspartate, 155
Stearoyl ethyl ester, 187
Stearoyl serine methyl ester, 151
Stearoylcysteine methyl esters, 13
STEM = scanning transmission electron microscopy, 153
Stereochemistry, 76
Steric force(s), 21, 98
Steric repulsion, 164
Steroid, 76, 140
Sticky ends, 146
Stiffening, 96
Stilbene, 92
STM, *see* STEM
Stoppers for pores, 6, 81–83
Streptavidin, 156
Subcell, 188
Sucrose, 75
Succinyl amphiphiles, 12, 18, 43, 50–58
Sulfide oxidation, 19, 47
Sulfonate salt, 161
p-Sulfonatophenyl, 139
Sulfoxide, 47
Sunlight, 7
Supramolecular assembling, 117
Surface energies, 122
Surface monolayer, 149
  definition, synkinesis, 3
Surface pressure, 151
Surfaces, creation, 4
Swelling of lecithin, 99
Synkinesis, 2–4, 173, 182, *see* Synkinon
Synkinetic domain, 76–83
Synkinon, 2, 4–6, 7, 11, 98, 104–106, 141
  amide, 7
  amphiphile, 7
  asymmetric, 7–8, 14
  bolaamphiphile, 11, 50–58, 86, 175–181
  carbohydrate, 7

Subject Index

cast films, 172
cationic, 11
chromatography, 7
co-crystals, 207–212
crystals, 186, 207
D,L-cyclopeptide, 141
definition, 2–5
diazotized, 11
facial, 34
ferrocene, 17, 39
fibres, 98–99
flat cyclopeptide, 141
fluid threads, 102–104
fluorescent dyes, 6
fluorinated, 7
glutamic acid, 7
hairy rods, 173
macrolide, 13
micelle, 28–29
multilayers, 172–174
nanopore, 165–170
oxidative coupling, 10–11
polymerizable, 11
pores, 75–76, 80–84, 165–170
porphyrins, 69–73, 126–139
reactive, 8, 10
ribbons, 110
rigid, 34–36
rods, 114–126, 147
synthesis, 7–19
scheme, 5
shape, 186
solid state, 182
solubility, 185
spacer lipid, 11, 17
steroid, 34–36, 68, 95
stoppers, 6, 81–83
structural components, 4–6
synthesis, 7–19
target assemblies: domains, pores, edge amphiphiles, 4
triple layer, 106–113, 141–142, 156–157
tubules, 106–113, 141–142
vesicles, 50–69

Tail-to-tail bilayer, 118, 187–170
Tailored excitation pulse, 32
Tape structure, 210
Tartaric acid amides, 13, 193
TBS = trinitrobenzenesulfonic acid, 93

Tecton, 208
TEM = transmission electron microscopy, *see* electron microscopy
Tetracyanoquinodimethane, 178
Tetraalkylammonium, 8, 11, 81
Tetrahedral network, 208
Tetramethyl benzidine, 33
Tetraphenylborate, 93–95
Tetrasulfonated aniline, 45
Tetrathiafulvalene, 178
Thiobis(ethyl acetoacetate), 163
Thioglycosides, 14–15
Thiolate, 77, 160–161
Thiophenolate, 77
Thorium(IV), 129
Threading, 140
Threonine, 151
Thyminyloctyl ω-ammonium salt, 30
Tilt angle, 161
Tin oxide glass electrode, 165
Tin(IV) porphyrinate, 131–132
TPP = *meso*-tetraphenylporphyrin 72, 127, 128, 134, 200, 211
Trans-cinnamic acid, 32
Transphosphatidylation, 8–9, 11
Trapping of fibres, 117
Triad pattern, 210
Triple layer, 157
Tryptophan, 43
Tubes or tubules, *see* Fibrous assemblies

U-shaped conformer, 190
Ultrasonication, 87
UMP = uracil monophosphate, 155
Undulation, 54
Uranium(IV), 129
Urea, 208
UV = ultraviolet, 155, 205
  excitation, 161
  irradiation, 19, 13
UV/VIS spectrum, 179

Van der Waals interactions, 21, 193
Vanadium porphyrin, 137
Vaska-type iridium compound, 147
Vaterite, 152
Vesicle, 3, 49–97, 106–113
  acceptor head groups, 49
  acridine orange, 56–57

Vesicle (*cont.*)
   aggregation, 54
   aggregation number, 49
   amino acid transport, 84
   amino acids, 87
   asymmetric, 3, 50, 56, 58
   azobenzene indicator, 60
   BLM, asymmetric, 59
   budding, 96
   butadiyne unit, 88
   carbohydrate recognition, 63
   carotenoids, 73–75
   cation–proton transport, 80
   chain mobility, 51
   chiral surface, 49
   chlorophyll, 70
   cis–trans isomerization, 92
   cmc = critical micellar/vesicular
      concentration, 39, 49, 50
   coated, 91
   compartments, 66
   conversion to tubules, 96
   counterion entrapment, 55
   covesicles, 59
   curvature, 50
   definition, 3, 49
   DHP, 70
   diazonium photodecomposition,
      59–60, 62
   dihexadecyl polycationic, 89
   diyne, 89
   DNA entrapment, 85–86
   DODAB, 70
   domains, 75, 79
   donor head groups, 49
   DPPC, 66, 75
   DSC, 87
   EDTA, 60
   electron microscopy, 89
   electron transfer, 70
   entrapment, 49, 85
   exo/endo, 77
   flattening, 94
   flip-flop, 58–62, 92
   formation
      by half-reduction of bis-viologen, 37,
         56, 62
      from micelles, 38
      by one-sided precipitation, 37, 55, 58
      by protonation of soaps, 53
   fusion, 94–97
   giant, 96
   glutamic acid, 106
   glutathione, 70
   glycose tetraacetates, 63, 65
   growth, 96
   H-aggregate, 61
   haemagglutinin, 65
   harpoon, membrane-disrupting, 76
   head group surface, 53
   hydrophilic spacer, 87
   ion channel, 79–84
   ion transport, 75–84
   lectins, 63–65
   lyophilization, 99
   membrane proteins, 85
   membrane thickness, 50–51
   metal ions, 60–61, 75–89
   microvesicle, 66
   MLM stability, 59
   molecular wire, 72
   monomers in equilibrium, 49
   name, 49
   NMR, 85
   one-sided precipitation, 55
   oxidative polymerization, 98
   pH gradient, 77, 80
   phase transition, 66, 89
   phosphatidylcholine, 73
   photoreaction, 84, 92–95
   photospiran, 84
   pK_a value, ylide, 60–61
   polymerization, 39, 85
   pores, 75, 79–84
   porphyrin, 69–73
   quinone, 70
   receptor, 79
   repulsion, 96
   ripple structure, 66
   rupture, 92
   skeletonized, 89
   smallest, 54
   stereochemistry, 76
   stiffening, 96
   stoppers for ion channels, 81–83
   surface reaction, 59–62
   synkinesis, 3
   tubular, 99, 106–113
   undulation, 49, 54
   voids, 50

Vesicle (*cont.*)
  volume, 49
Viologen, 69, 164, 179
Vitamin K_3, 43
Vitreous ice, 29, 103
Void, 50

Washing, 44
Water
  degradation by light, 33
  hydration layers, 22–23
Water-soluble porphyrin, 174
Wax cuticula, 99

Wedge-shaped "molecular harpoon", 76
Wetness, 27
Wilkinson catalyst, 8
Wire, 6, 144

XPS = X-ray photoelectron spectroscopy, 161, 190

Zeolite, 45
Zipper reaction, 12
Zinc formylbiliverdinate, 144
Zinc porphyrinate, 72, 127, 128, 130, 133